Actes du XIVème Congrès UISPP, Université de Liège, Belgique, 2-8 septembre 2001

Acts of the XIVth UISPP Congress, University of Liège, Belgium, 2-8 September 2001

I0093201

SECTION 1 : THÉORIES ET MÉTHODES / THEORY AND METHOD

Colloque / Symposium 1.7

Three-Dimensional Imaging in Paleoanthropology and Prehistoric Archaeology

Edited by

Bertrand Mafart
Hervé Delingette

With the collaboration of

Gérard Subsol

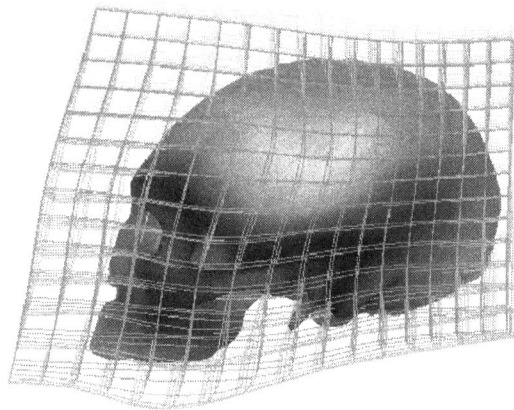

BAR International Series 1049
2002

Published in 2016 by
BAR Publishing, Oxford

BAR International Series 1049

Acts of the XIVth UISPP Congress, University of Liège, Belgium, 2-8 September 2001
Colloque / Symposium 1.7

Three-Dimensional Imaging in Paleoanthropology and Prehistoric Archaeology

ISBN 978 1 84171 430 1

© The editors and contributors severally and the Publisher 2002

Typesetting and layout: Darko Jerko

BAR Publishing is the trading name of British Archaeological Reports (Oxford) Ltd.
British Archaeological Reports was first incorporated in 1974 to publish the BAR
Series, International and British. In 1992 Hadrian Books Ltd became part of the BAR
group. This volume was originally published by Archaeopress in conjunction with
British Archaeological Reports (Oxford) Ltd / Hadrian Books Ltd, the Series principal
publisher, in 2002. This present volume is published by BAR Publishing, 2016.

Printed in England

BAR
PUBLISHING

BAR titles are available from:

BAR Publishing
122 Banbury Rd, Oxford, OX2 7BP, UK
EMAIL info@barpublishing.com
PHONE +44 (0)1865 310431
FAX +44 (0)1865 316916
www.barpublishing.com

TABLE DES MATIÈRES

3D IMAGING IN PALEOANTHROPOLOGY AND PREHISTORIC ARCHEOLOGY: A NEW TOOL FOR OLD SCIENCES OR AN EMERGING SCIENCE?

Bertrand MAFART

The program of the XIVth UISPP Congress included a colloquium entitled "3D Imaging in Paleoanthropology and Prehistoric Archeology" coordinated by Hervé Delingette and myself with the active collaboration of Gérard Subsol. This colloquium provided an unprecedented forum for presentation of high-quality studies and for instructive discussion. I would like to take this opportunity to express my gratitude to the 16 speakers from 9 countries who participated.

I fully expect that there will be a strong demand for these Colloquium Proceedings. This volume constitutes the most extensive collection of data yet published on the topic. In addition to results documenting the value of 3D imaging in a wide range of applications, researchers and students will find a raft of useful information including references, methodologies, and analysis techniques.

Paleoanthropologists and archeologists have always depended on the mind's "natural" 3D imaging ability to analyze material uncovered during excavations. Man's vision is stereoscopic and the neuronal connections in his brain create detailed virtual representations. Mental images are held in memory and can be reproduced verbally or pictorially. The great advantage of computer-based technologies is to automate data collection and image reconstruction. An additional benefit is to allow limitless re-processing and transfer between users.

Despite these advantages many investigators still express reticent to the introduction of 3D imaging and virtual reality techniques. Many still wonder what these techniques actually add to traditional techniques. This question is usually followed by objections that visual examination and description are more precise than digital imaging. The colloquium proceedings contain answers to such questions and objections.

3D images have had a great impact on paleoanthropology. Visualization of intraosseus cavities such as cranial sinuses and semi-circular channels has opened the way to analysis of once invisible structures. Certainly the most clear-cut benefit of 3D imaging techniques involves morphometric analysis. There also is an obvious need for computer assistance spatial analysis of the massive amount of fossil material collected at prehistoric excavation sites. New techniques for automation of 3D data acquisition and spatial positioning hold the promise of allowing reconstitution of the soil of prehistoric settlements and of 3D modeling of the geological and archeological dynamics of the site.

Another reason that 3D imaging that has been beneficial is that it brings together teams of investigators with wide-array of expertise. In paleoanthropology, 3D imaging requires a radiologist for acquisition of scan images, a computer scientists for image processing, and a paleoanthropologist for coordination of anthropological analysis. In prehistoric archaeology, image acquisition often requires a skilful technician able to adapt industrial surface scanning techniques to archeological soils or objects, the services of a computer technician, and the cooperation of the team in charge of the archeological site. The level of technical complexity is such that no investigator can have the full range of competency and yet each must know and understand the problems of the other players. Perhaps for this reason a sort of intellectual alchemy took place between diverse groups of investigators who participated in the colloquium.

Every emerging field has its pioneers. For 3D imaging of hominid fossils it was Franz Zonneveld who started 16 years ago. In his presentation at the colloquium, F. Zonneveld described methodological issues in fossil scanning such as variable degrees of mineralization, inclusion in a matrix, and image resolution. He also reviewed current techniques and above all future perspectives of 3D imaging.

Analysis of sinus cavities and cranial pneumatization has benefited greatly from the advent of three-dimensional reconstruction based on computed tomography data (3D-CT). T. C. Rae and T. Koppe showed how indispensable these techniques have become for the study of non-human primates and mammals in general. Based on their study of paranasal pneumatization in catarrhine primates with special focus on the maxillary sinus, these authors were able to revise previous interpretations of growth and evolution of sinuses in apes and old-world monkeys.

Assessment of endocranial casts is often difficult. The main benefits of 3D imaging are to allow virtual isolation of bones from the geological matrix and production of accurate 3D-hardcopies by stereolithographic modeling. E. Bruner, G. Manzi, and P. Passarello studied the virtual endocast of the Neanderthal Saccopastore 1 specimen and used geometric morphometric analysis to compare results with hominid fossil specimens from the middle and late Pleistocene period. In their conclusion these authors stated that variability in endocranial morphology within the genus Homo appeared to be strongly correlated with the size and expansion of parietal areas.

3D studies have benefited greatly from advances in geometric morphometrics. Up to now, paleoanthropological analysis of the particularly complex morphology of the temporal bone has been difficult. Based on 15 landmarks identified by 3D imaging on the temporal bone, K. Harvati were able to compare differences between Neanderthals and modern humans and between two species of chimpanzees using the Procruste method. Findings indicated that Neanderthals differed more from modern humans than the two chimpanzee species differed from each other.

Medical computed tomography is not the only method allowing 3D image acquisition. M. Friess, L. F. Marcus, D. P. Reddy, and E. Delson used laser surface scanning to assess the relative surface areas of modern hominid skull specimens from various geographical locations and fossilized specimens and to re-appraise theories that variations in the facial morphology of Homo sapiens sapiens are related to cold adaptation. Their findings suggested that facial morphology in Inuit or Neanderthal populations was unrelated to climatic conditions.

Current 3D morphometric techniques do not allow complete analysis of fossil specimens. G. Subsol, B. Mafart, A. Silvestre and M.-A. de Lumley presented a computer-assisted technique that allows 3D visualization and analysis of CT scan images of fossils as well as comparison of these images with each other and with images from modern humans. Homology points between skulls were determined automatically. Future applications in facial reconstruction and three-dimensional morphometry were also presented.

Despite its importance for paleoanatomical analysis of fossil morphology, time is usually factored in after spatial analysis. C.P. Zollikofer and M.S. Ponce De León studied the influence of this veritable "fourth" dimension that has three distinct yet interconnected aspects, namely ontogeny (individual development), phylogeny (speciation) and diagenesis (fossilization). These investigators described several computer-assisted models of geometric morphometric analysis that take into account the effects of these three aspects of the temporal dimension on fossil morphology.

Currently 3D imaging techniques are systematically used in conjunction with traditional analytical methods of fossil analysis. To demonstrate the synergy between these two techniques, J. L. Thompson, A. J. Nelson and B. Illerhaus re-examined the Neanderthal skull of the Moustier 1 adolescent. In addition to studying facial sinuses and determining cranial volume, they were able to obtain a virtual reconstitution of the fossil. Use of the two methods led to a better understanding of the ontogenic and phylogenetic features of this fossil.

3D imaging using computer tomography, magnetic resonance imaging, or surface laser scanning allows "virtual reality" representation of fossils. Specimens can be studied and even reconstructed repeatedly without altering the original. After reviewing the advantages of computerized techniques of data processing and imaging, G. Weber addressed the issue of availability to the scientific community at large. The author advocated creation of global access to a digital 3D-data archive of recent and fossil hominoids.

One of the major challenges of 3D imaging is development of a reliable method of facial reconstruction. Such techniques could be used not only for identification of missing persons

from skeletal remains in forensic medicine but also for picturing prehistoric men in archeological studies. G. Odin, G. Quatrehomme, G. Subsol, H. Delingette, B. Mafart and Marie-Antoinette de Lumley compared the results of manual and computerized-assisted techniques for 3D reconstruction of the face of the Tautavel Man. Based on their findings, these investigators concluded that the two techniques are complementary.

Tooth pattern analysis is now within the scope of 3D imaging. O. Kullmer, M. Huck, K. Engel, F. Schrenk and T. Bromage described the use a portable 3D optical topometry system to achieve high-resolution tooth images in an early hominid from Java, Indonesia. The technique allowed virtual modeling of the occlusional surface and morphometric analysis. These data along virtual reconstructions and animations were placed in an image bank.

In addition to static analysis of tooth patterns, 3D imaging allows simulation of dynamic changes due to abrasion. I. L. Gügel and K.-H. Kunzelmann described an experimental simulator designed to quantify dental abrasion resulting during chewing of different cereal species. 3D laser imaging of enamel surfaces were obtained before and after chewing simulation. The findings of this study confirm the efficacy of 3D imaging methods for assessment of the mechanisms underlying dental abrasion.

Archeology is one of the fields offering the widest range of applications for 3D imaging techniques. This point was well illustrated in several colloquiums entitled « Computer Applications and Quantitative Methods in Archaeology ». Pre-historians have already reaped many benefits from computerized data management. Advantages include easier retrieval and processing of data, spatial representation of archeological specimens, and simulation of excavation sites and their geological evolution.

The strengths and weaknesses of current techniques for digital representation of archeological datasets were described by H. Delingette. The main advantages of an entirely digital archeological database are conservation in time and space, mass data handling, and computation of objective measurements. The authors also described current methods of geometric modeling of archeological specimens or sites. A distinction was made between the acquisition, modeling and editing phases of digital processing.

Processing and analysis of archaeological shards is both time-consuming and labor-intensive. R. Sablatnig, S. Tosovic and M. Kampel described a computer-assisted documentation system in which classification and semi-automated reconstruction is based on 3D representations. The ultimate objective of this system will be to automate the tasks of archivage and 3D acquisition for archeologists.

The results of multidisciplinary studies in the Middle Pleistocene cave site in Arago have been placed in a database entitled «Prehistoric and Paleontologic Material» or scanned into a digital library. H. de Lumley, C. Butour, A.-M. Moigne, V. Pois and R. Vaudron presented a 3D reconstruction of the prehistoric settlement and a simulation of the geodynamic evolution of quaternary filling of the Caune de l'Arago.

The studies presented at this colloquium show that 3D imaging techniques do not compete with or detract from other methods. As their availability and power grows, computerized systems based on technologies with capabilities far beyond those of the human eye (e.g. industrial scanners and high-resolution laser) will become indispensable tools alongside classic anatomical and morphological analysis. This technology should not be considered as an emerging science in itself but rather as a powerful new multidisciplinary tool to assist paleoanthropological and archeological studies.

Development of 3D imaging for paleoanthropology and prehistoric archeology will be as great over the next decade as that of the Internet for communications was over the last ten years. Computer systems allow limitless storage, analysis, and exchange of data. Geometric morphometry with 3D images will demonstrate the full potential of analysis. This technology is fast becoming an integral part of the scientific methodology both for research and reporting. Soon no investigator will be able to avoid using these techniques in his work. Research centers need to have a clear policy to promote and encourage this development for the greater benefit of pre-historical and protohistoric sciences.

Adress of the author:

Laboratoire d'Anthropologie
Faculté de Médecine secteur Nord
Boulevard Pierre Dramard
13916 Marseille Cedex 20
bmafart@aol.com

APPLICATIONS AND PITFALLS OF CT-BASED 3-D IMAGING OF HOMINID FOSSILS

Franz ZONNEVELD

Résumé: La mise au point de scanner de haute résolution pour la médecine, à partir de 1980 a permis d'une part de visualiser les structures internes du corps humain, d'autre part d'en obtenir des reconstitutions tridimensionnelles. Des problèmes méthodologiques importants ont été rencontrés avant de pouvoir obtenir des images de qualité et doivent être bien connus. L'auteur explique à partir d'exemples, les raisons de ces difficultés, spécifiquement liées aux particularités des matrices minérales et les solutions trouvées pour les résoudre. Il expose, en les illustrant par des exemples, les nombreuses nouvelles possibilités d'analyses offertes par l'imagerie scanner tridimensionnelle encore trop peu employée.

Abstract: The advent of high-resolution medical computed tomography, became possible to visualize the internal structures of hominid fossils, to obtain 3D-images. The author review five common technical pitfalls that may adversely influence the cross-sectional CT and 3D-images reconstructed. He explains the different solutions for each problem. Then, he present some example of applications of CT-based 3-D imaging in paleoanthropology as disarticulation of the different constituents of a fossil, most commonly bone and matrix, quantitative analysis, reconstruction of fossils in a computer using both internal and external landmarks and techniques such as mirror imaging for filling in missing parts, use of segmented 3-D data sets to physical models for visual inspection, teaching create.

INTRODUCTION

Around 1980, with the advent of high-resolution medical computed tomography (CT) for the examination of the internal morphology of patients, it became possible to visualize the internal morphology of hominid fossils as well (Wind, 1984), (Zonneveld & Wind, 1985). This was an excellent alternative to the radiography of dense and matrix-filled fossils (Wind & Zonneveld, 1985). Around the same time frame, three-dimensional medical imaging, based on a volume of CT-slices, was also being developed (Hemmy *et al.*, 1994). It thus became available, although still of primitive quality, for use in paleontological (Conroy & Vannier, 1984) and palaeoanthropological studies (Wind & Zonneveld, 1989), (Conroy, Vannier & Tobias, 1990). Over time, the technique of 3D-imaging improved drastically and it became possible to segment and disarticulate specific structures separately and combine them in a single 3D-image making use of different colors (Zonneveld, Spoor & Wind, 1989). Here I briefly review some common technical pitfalls and applications of CT-based 3-D imaging in palaeoanthropology. For recent, more comprehensive reviews see, among others, Zollikofer, Ponce de León & Martin (1998), and Spoor, Jeffery & Zonneveld (2000*a*&*b*).

CT PITFALLS

There are a number of pitfalls in CT that may adversely influence both cross-sectional CT as well as the 3D-images reconstructed therefrom. Table 1 lists both the causes as well as the results of these deficiencies.

Pitfall 1.

If a fossil is dense (e.g. due to mineralization), or if it has a large dimension in certain directions, it may happen that the standard deviation of the noise in the raw data is in the same order of magnitude as the signal itself.

This signal results from the radiation passing through the object at the moment it is detected by the X-ray detector.

Table 1. Causes and their resulting imperfections in the CT image.

Lack of signal in X-ray beam	"Frozen noise" in CT scan
Scanning of too high density in object	White overflow
Scanning of too low density in object	Black overflow
Lack of beam hardening correction	Image inhomogeneity
Partial volume averaging in thick slices	Artifacts, smoothing

In such a case, the noise in the raw data may be "frozen" into the reconstruction of the CT slice (Spoor, Jeffery & Zonneveld, 2000a). The effect on 3D-imaging will be that it is virtually impossible to perform object segmentations in the region of the "frozen noise" and the resulting object surface in the 3-D image will therefore have a "mottled" appearance due to the lack of object surface definition.

Pitfalls 2 and 3.

The CT number scale (Hounsfield Scale) is limited to a certain density range (this is usually -1000 to +3095 HU) whereby water is at 0 HU and air at -1000 HU. Sometimes a fossil is too dense. For example, this has been encountered in the Broken Hill fossils which are impregnated with lead- and zinc-containing minerals (Zonneveld & Wind, 1985), but is commonly found in other types of fossilization as well. Similar problems may occur when a fossil is small and reasonably dense. In that case, a lack of beam hardening can play a role. Beam hardening causes the effective energy of the radiation to raise as low-energy beams are being attenuated more severely than high-energy beams. The higher the beam energy the lower the CT number produced, and vice versa. Due to the lack of beam hardening in small fossils the resulting CT-numbers may be extremely high such as in tooth enamel when scanning single teeth (Spoor, Zonneveld & Macho, 1993). In all of these cases it may happen that part of the object cannot be fitted within the CT number scale. In pitfall 2, all tissues outside the range will be shown as white. I call this "white overflow". In practice, this means that, in 3D-imaging, the interface between the tissue within the CT number range and that outside of this range will be displaced which may result in wrong thickness and wrong volume (e.g. in case of enamel thickness and volume measurements). In pitfall 3, all tissues outside the range will be usually shown as black as the value 4096 will be subtracted from their true CT number lying outside the normal range. This means that the parts of the object that are too dense show dropouts. In 3D imaging, an object that is partly too dense can still be segmented, but underneath its surface there may be only a very thin layer of tissue within the normal range, the remainder is too low as a result of the dropout. That means two things: if the normal layer is too thin, the 3D-image will show holes, and if the volume has to be determined, the black overflow zone does not contribute to that volume resulting in volumes that are significantly too low. To avoid pitfalls 2 and 3 it is advantageous to use a CT scanner with the capability of an extended CT-number scale. In case of scanning isolated teeth, it can be helpful to surround the fossil by a radiation absorbing medium, e.g. thick Plexiglas cylinder. This will avoid the high CT-numbers in the fossil tooth.

Pitfall 4.

It is the result of the fact that CT scanners are usually not calibrated for fossils. As the beam hardening is more severe in fossils due to the mineralization, the reconstructed CT-numbers will deviate from the real numbers and will start to show a relationship with the position of the tissue in the object. Thus deeply lying tissues get lower CT-numbers and tissues just under the surface get higher CT-numbers. This creates an unintended inhomogeneity in the CT image and creates unintended interfaces in segmentation for 3D-imaging. The only solution for this problem is to recalibrate the CT scanner for fossils e.g. by means of aluminum phantoms instead of Plexiglas ones (Zonneveld & Wind, 1985).

Pitfall 5.

It is partial volume averaging. This means that a kind of averaging occurs within the volume elements (voxels) represented by the image elements (pixels) in the final image. Such an averaging effect is most severe when the slice thickness used during CT scanning is relatively thick. Sometimes such a thick slice is required to avoid frozen noise. Partial volume averaging may have two effects. The first effect is image artifacts due to the fact that during the image reconstruction a logarithm is taken. The mixing of signals before taking this logarithm is the source of the artifact as the mixing should actually take place after taking the logarithm. The most severe artifacts occur when high (mineral) and low densities (air pockets) are mixed in a single slice. The second effect is a smoothing effect due to the loss of detail that can be resolved by a thicker slice. In 3-D images the result is a displacement of the true interface between structures and a lack of detail.

SPECIFIC APPLICATIONS OF 3D-IMAGING

A *first application* in 3D-imaging is the disarticulation of the different constituents of a fossil, most commonly bone and matrix, but also plaster, and resin, if these have been used in the reconstruction of the specimen. This disarticulation enables not only the visualization of the different constituents of the fossil but also allows for further processing of the individual components such as volume calculation and model fabrication. For example, using this approach, it was possible to visualize the SK 47 *Paranthropus robustus* cranial base without the endocranial matrix thus demonstrating a groove of a right occipital-marginal sinus (Spoor & Zonneveld, 1999). Figure 1 shows the Neanderthal cranium Tabun C1, visualized

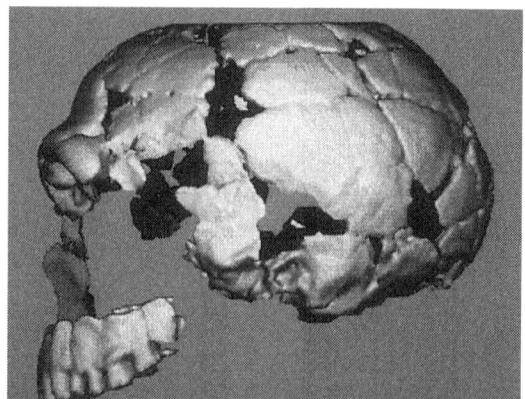

Figure1 - Representation of the bony fragments of the Tabun C1 cranium. This left lateral view is a 3-D reconstruction based on CT scans while the plaster between the bony fragments has been left out.

without the plaster used in the physical reconstruction, thus demonstrating how it is composed of a large number of bone fragments. Furthermore, Figure 2 demonstrates the dental root configuration of the Mauer 1 mandible.

A *second application* in 3D-imaging is its use in quantitative analysis. Examples are the calculation of volumes of the endocranial cavity (Figure 3) and the paranasal sinuses. (E.g. Conroy et al., 1998) (Spoor & Zonneveld, 1999), and as a source of 3-D morphometric landmarks (e.g. Spoor *et al.*, 1999) (Ponce de León & Zollikofer, 2001).

A *third application* is the reconstruction of fossils in a computer using both internal and external landmarks and techniques such as mirror imaging for filling in missing parts (Zollikofer *et al.*, 1995). For example, mirror imaging of the better-preserved side of the Sangiran 4 Homo erectus maxilla provides an improved image of its dental arcade and palate

(Figure 4). This third application will be discussed in a separate contribution as it was presented separately during the meeting by CPE Zollikofer (Zollikofer & Ponce de León, 2002).

A *fourth application* is the use of segmented 3-D data sets to create physical models for visual inspection, teaching etc. Today, such models are usually made using the technique of stereolithography (Zollikofer & Ponce de León, 1995), and may represent the fossils as one finds them in museums (Weber, 2001) or fossil reconstructions as described under the third application. Other possible rapid prototyping techniques are the milling of a model from PUR foam (Zonneveld & Noorman van der Dussen, 1992), and fused deposition modeling (FDM) using wax or a plastic called ABS. This fourth application will also be discussed in a separate contribution as H. Seidler (2002) discussed it in a separate presentation during the meeting.

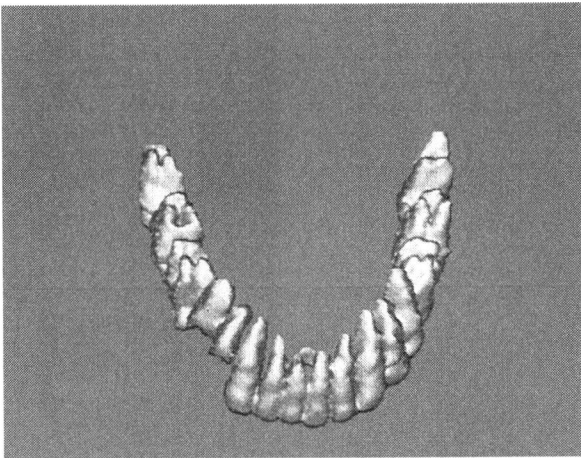

Figure 2 - Antero-inferior view of the root configuration of the dentition in the Mauer 1 mandible by leaving out the mandibular bone.

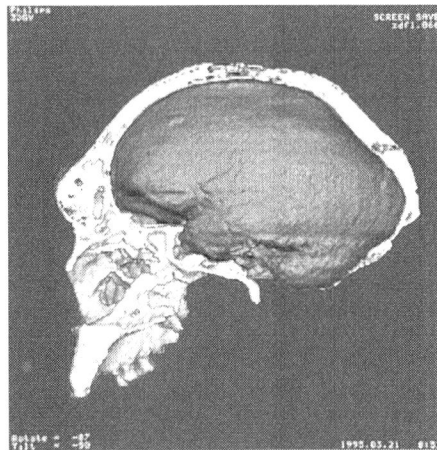

Figure 3 - Visualization of the segmented left endocranium of the Broken Hill 1 cranium using a selective cut-away view eliminating the bone of the left cranial half.

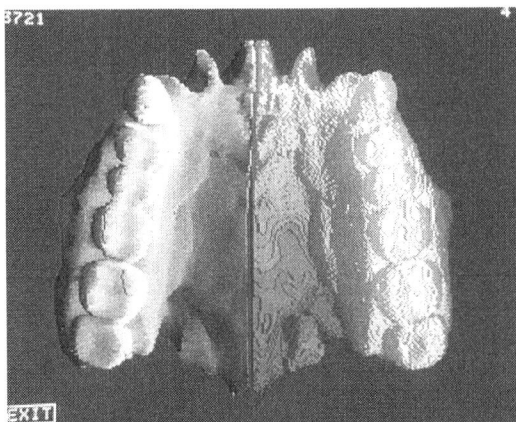

4a. Original fossil of the maxilla of Sangiran 4 showing distortion of the left hemi-maxilla

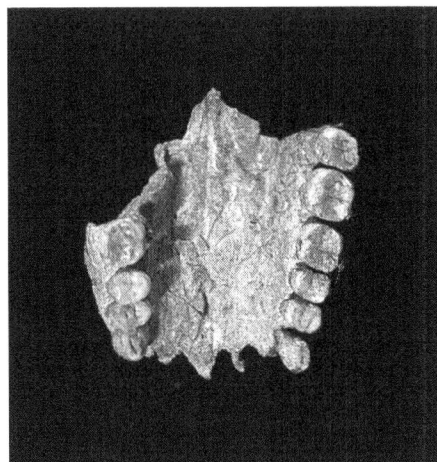

4b. Reconstruction of the most probable original shape of this fossil by means of mirror imaging.

Figure 4 - Fossil reconstruction my means of mirror imaging.

A derived version of the fourth application, which has a limited scientific value due to lacking information, is the creation of an individual facial reconstruction. This is an old technique originated by Gerassimov (1968), who reconstructed faces, using casts, of several hominid fossils (Sts 5, Sangiran 4, Peking man, Mauer, Steinheim, Le Moustier 1, Cro-Magnon, Combe-Capelle, La Quina 5, Saccopastore 1, Broken Hill 1, Gibraltar 1 and many others). R. Neave of the University of Manchester has modernized this technique which can now also be applied using the physical models described above (Prag & Neave, 1997).

My own experience is limited to the reconstruction of a mummy's face (Sensaos) (Raven, 1998) (Figure 5) and that of a living person to test the quality of the technique. R. Neave did both facial reconstructions. I have no experience with the application of this technique to hominid fossils.

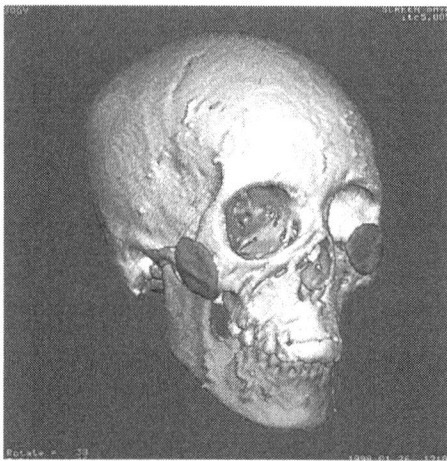

5a. 3-D image showing the displaced artificial eyes.

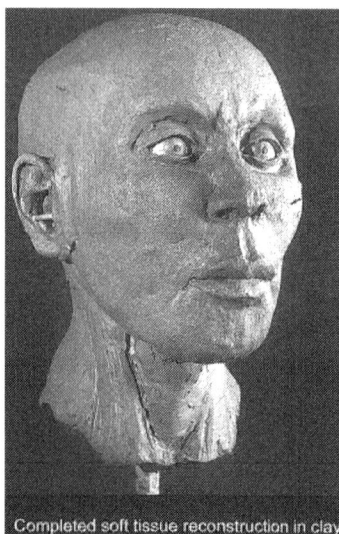

5b. Reconstruction of Sensaos' face with clay on top of the model derived from the segmented CT-data (Reconstruction by Richard Neave, Manchester University, U.K.).

Figure 5 - Facial reconstruction of a mummy's face (Sensaos).

Acknowledgements

The author is grateful to C.F. Spoor, Ph.D. for revision of the manuscript, as well as to D. Dean, Ph.D., Case Western Reserve University Medical School, Cleveland, Ohio, U.S.A. and M. Raven, Ph.D., Egyptologist, State Museum of Antiquities, Leiden, the Netherlands for permission to publish Figs. 4a and 5b respectively.

BIBLIOGRAPHY

CONROY, G.C., & VANNIER, M.W., 1984, Noninvasive three-Dimensional computer imaging of matrix-filled fossil skulls by high- resolution Computed Tomography. *Science* 226, 456-458.

CONROY, C.G., WEBER, G.W., SEIDLER, H., TOBIAS, P.V., KANE, A., BRUNSDEN, B. ,1998, Endocranial capacity in an early hominid cranium from Sterkfontein, South Africa. *Science* 280, 1730-1731.

CONROY, G.C., VANNIER, M.W. & TOBIAS, P.V. ,1990, Endocranial features of *Australopithecus africanus* revealed by 2- and 3-D Computed Tomography. *Science* 247, 838-841.

GERASSIMOW, M.M. (1968). *Ich suchte Gesichter.* Gütersloh: C. Bertelsmann Verlag.

HEMMY, D.C., ZONNEVELD, F.W., LOBREGT, S., & FUKUTA, K., 1994, A decade of clinical three-dimensional imaging: a review. Part 1. Historical development. *Invest Radiol* 29, 489-496.

PONCE DE LEÓN, M.S. & ZOLLIKOFER, C.P.E., 2001, Neanderthal cranial ontogeny and its implications for late hominid diversity. *Nature* 412, 534-538.

PRAG, J. & NEAVE, R. , 1997, *Making faces. Using forensic and archaeological evidence.* London: British Museum Press.

RAVEN, M.J., 1998, Giving a face to the mummy of Sensaos in Leiden. *KMT* 9 issue 2, 18-25.

SEIDLER, H., 2002, (personnal communication, during UISPP satellite colloquium"Applications of three-dimensional informatics methods to human paleontology and prehistoric archeology".

SPOOR, C.F., ZONNEVELD, F.W. & MACHO, G.A. ,1993, Linear measurements of cortical bone and dental enamel by Computed Tomography: Applications and problems. *Am J Phys Anthrop* 91, 469-484.

SPOOR, F. & ZONNEVELD, F. 1999, Computed Tomography-based three-dimensional imaging of hominid fossils: features of the Broken Hill 1, Wadjak 1, and SK 47 crania. In (T. Koppe, H. Nagai, & K.W. Alt, Eds) *The paranasal sinuses of higher primates. Development, function, and evolution*, pp. 207-226. Chicago: Quintessence Publishing Co, Inc.

SPOOR, F., O'HIGGINS, P., DEAN, C. & LIEBERMAN, D.E.., 1999, Anterior sphenoid in modern humans. *Nature* 397, 572.

SPOOR, F., JEFFERY, N., & ZONNEVELD, F. , 2000a, Imaging skeletal growth and evolution. In (P. O'Higgins & M. Cohn, Eds) *Development, growth and evolution*, pp. 123-161. London: Academic Press.

SPOOR, F., JEFFERY, N. & ZONNEVELD, F. , 2000b, Using diagnostic radiology in human evolutionary studies. *J. Anatomy* 197, 61-76.

WEBER, G.H. ,2001, Virtual anthropology (VA): A call for *Glasnost* in Paleoanthropology. *Anat Rec (New Anat)* 265, 193-201.

WIND, J. ,1984, Computerized X-ray tomography of fossil hominid skulls. *Am J Phys Anthrop* 63, 265-282.

WIND, J. & ZONNEVELD, F.W. ,1985, Radiology of fossil hominid skulls. In (P.V. Tobias, Ed) *The past, present and future of hominid evolutionary studies*, pp. 437-442. New York: Alan Liss.

WIND, J. & ZONNEVELD, F.W. ,1989, Computed tomography of an *Australopithecus* skull (Mrs Pless): a new technique. *Naturwissenschaften* 76, 325-327.

ZOLLIKOFER, C.P.E. & PONCE DE LEÓN, M.S. ,1995, Tools for rapid prototyping in the biosciences. *IEEE Comp Graph APPL* 15, ISSUE 6, 48-55.

ZOLLIKOFER, C.P.E., PONCE DE LEÓN, M.S., MARTIN, R.D., & STUCKI, P. ,1995, Neanderthal computer skulls. Nature 375, 283-285.

ZOLLIKOFER, C.P.E. PONCE DE LEÓN, M.S. & MARTIN, R.D, 1998, Computer- assisted paleoanthropology. *Evolutionary Anthropology 6*, 41-54.

ZOLLIKOFER, C.P.E. & PONCE DE LEÓN, M.S., 2002, virtual paleoanthropology: the 4th dimension In UISPP proceedings of the colloquium"Applications of three-dimensional informatics methods to human paleontology and prehistoric archeology", Archeopress.

ZONNEVELD, F.W. & WIND, J., 1985, High resolution computed tomography of fossil hominid skulls: a new method and some results. In (P.V. Tobias, Ed) *The past, present and future of hominid evolutionary studies*, pp. 427-436. New York: Alan Liss.

ZONNEVELD, F.W., SPOOR, C.F. & WIND, J.,1989, The use of CT in the study of the internal morphology of hominid fossils. *Medicamundi* 34, 117-128.

ZONNEVELD, F.W., NOORMAN VAN DER DUSSEN, M.F. 1992, Three-dimensional imaging and model fabrication in oral and maxillofacial surgery. *Oral Maxillofac Surg Clin N Am* 4, 19-33.

3D IMAGING AND MEASUREMENT IN STUDIES OF CRANIAL PNEUMATIZATION

Todd C. RAE & Thomas KOPPE

Résumé: Les reconstitutions tridimensionnelles, à partir de coupes scannographiques, sont une source d'informations uniques et un outil aux remarquables potentialités pour étudier les structures internes du crâne. Désormais, il est possible de réaliser des études morphologiques détaillées de régions anatomiques mal connues auparavant, comme les cavités sinusiennes. Ainsi, les hypothèses précédemment émises concernant la croissance et l'évolution des structures internes crâniennes des mammifères peuvent être testées. Nous présentons ici l'étude par imagerie tridimensionnelle de la pneumatisation paranasale, en particulier des sinus maxillaires, chez les primates catarrhiniens. Cette étude a permis une analyse critique des interprétations précédentes de la croissance et de l'évolution des sinus chez les chimpanzés et les singes de l'Ancien Monde qu'il n'aurait pas été possible de réaliser sans l'apport, au plan de la visualisation et de la quantification, de l'imagerie tridimensionnelle à partir de données scannographiques.

Abstract: Three-dimensional reconstructions based on computed tomography (3D-CT) represents a unique and powerful tool for the study of internal structures of the skull. For the first time, detailed analyses of previously understudied features, such as cranial pneumatization, can be used to critically test hypotheses concerning the growth and evolution of aspects of the interior of the mammalian cranium. We report here on studies using 3D-CT to evaluate the expression of paranasal pneumatization, particularly the maxillary sinus, in catarrhine primates. As a whole, the work reported represents a critical re-evaluation of previous interpretations of the growth and evolution of the sinus in apes and Old World monkeys that would not have been possible without the aid of the visualization and quantification that 3D-CT provides.

INTRODUCTION

A persistent problem in cranial morphology is the characterization of internal structures, and particularly hollow spaces, such as sinuses. Increasingly, these morphological characteristics are used for functional and phylogenetic studies of mammals, driven in many cases by the discovery of fragmentary fossil evidence revealing previously unexamined aspects of craniofacial structure (Begun, 1992; Rae, 1999; Ward & Brown, 1986). Although gross differences and some basic measurements are available for these traits via destructive techniques, this is by no means ideal, particularly for rare or delicate specimens. Some non-destructive methods are also available (e.g., seed technique; see Shea, 1977), but in many cases these procedures are not applicable to both extant and extinct taxa, due to matrix infilling of fossil crania and increasingly strict rules on the use of seeds in museums.

Traditional radiography (e.g., Vlček, 1967; Blaney, 1986) circumvents some of the shortcomings of the techniques listed above, but is not without its own limitations. In many cases, fine details of shape are lost in ordinary x-ray visualization, and quantification is often restricted to a maximum dimension in a particular plane (Blaney, 2000). In addition, this technique can be sensitive to matrix infilling, confounding both measurement and interpretation of the internal structures of permineralized tissue.

Advances in medical imaging, however, have solved many of the problems facing those interested in internal cranial morphology. Virtual reconstructions in three dimensions, based on serial computed tomography (CT) scans (Figure 1), are an ideal way to derive shape and size data from structures inside the skull (e.g., Spoor & Zonneveld, 1999). Unlike laser-based methods, which can be used to study external structures only, 3D-CT combines the non-destructive internal visualization of traditional radiography with the ability to discern detailed shape data normally available only from invasive/destructive techniques (Koppe & Nagai, 1995). As a digital technology, 3D-CT also offers an unprecedented method for accurate measuring internal cranial structures, particularly 3D quantities, such as volumes of sinus spaces (Koppe, et al., 1996; Koppe & Nagai, 1999; Uchida, et al., 1998).

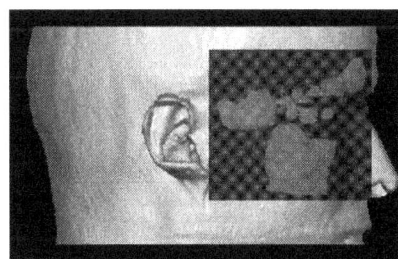

Figure 1 - Lateral view of three-dimensional virtual reconstruction from serial CT scans (3D-CT) of one of us (TCR). A semi-transparent window has been added to show 3D-CT reconstructions of the paranasal pneumatizations.

PRESENCE/ABSENCE OF INTERNAL STRUCTURES

Serial CT analysis, with or without subsequent 3D reconstruction, can be of great use in testing hypotheses of morphology. Even the simple discovery of the presence or absence of a particular pneumatization can alter our ideas about the evolution of characters in mammalian history. An example is taken from our recent work in the evolution of the maxillary sinus in Old World monkeys (Cercopithecoidea, Primates).

It has long been known (Paulli, 1900) that many cercopithecoids lack the maxillary sinus, a distinct pneumatic space in the cheek region of the face. The presence of this cavity in most other primates (Ward & Brown, 1986) and nearly all other eutherians (Moore, 1981) suggests that the sinus was present in the last common ancestor of placental mammals (Novacek, 1993). One Old World monkey, however, has been documented to possess a maxillary sinus; all known species of the genus *Macaca* possess a sinus space indistinguishable from that of other primates (Koppe & Ohkawa, 1999). Most workers have interpreted a small sinus to be the ancestral condition for the group (Harrison, 1987; Rae, 1997), although more recent analyses of extant taxa have suggested that the last common ancestor of the group may have lacked this pneumatic space (Rae, 1999). Nonetheless, this character state distribution among the extant forms makes the reconstruction of the ancestral state for the maxillary sinus in cercopithecoids equivocal.

One way to break the deadlock is to determine the character state in a stem taxon (*sensu* Ax, 1985) of the group. By acting as a closely related outgroup, a stem taxon (usually a fossil) can be instrumental in determining the polarity of a particular character state transformation, and thus provide a robust hypothesis of the ancestral state for the ingroup. In the present case, the only well-known stem cercopithecoid is *Victoriapithecus*, from middle Miocene (ca. 15 Ma) deposits of Kenya; character states of the dentition of this taxon suggest that it precedes the last common ancestor of the extant forms (Benefit, 1993), and this unique phylogenetic position has had a demonstrable impact on interpretations of craniofacial evolution in Catarrhini (Benefit & McCrossin, 1991, 1993). A recently discovered complete cranium of *Victoriapithecus* (Benefit & McCrossin, 1997) provides the perfect test case, as more fragmentary remains can be much more difficult to interpret. CT analysis was crucial in this instance, as the nasal cavity of the specimen was matrix-filled, making any visual inspection of the relevant areas impossible. The presence of heavily permineralized bone and copious matrix also precluded standard radiography.

The results of the CT analysis are unequivocal; *Victoriapithecus* does not possess a maxillary sinus (Rae, et al., in press). The lack of this pneumatization in the Miocene monkey (Fig. 2) strongly supports the interpretation of the absence of the sinus in the last common ancestor of extant cercopithecoids, which in turn implies that the corresponding pneumatization in the maxilla of the genus *Macaca* has evolved convergently (Rae, et al., in press). These results will undoubtedly have profound effects on interpretations of both the paleofunction and phylogeny of sinuses in primates. The discovery of the re-emergence of the maxillary sinus in *Macaca* may also help to clarify the (?functional) selection pressures responsible for the initial appearance of cranial pneumatization, which are unclear at present (Witmer, 1997, 1999).

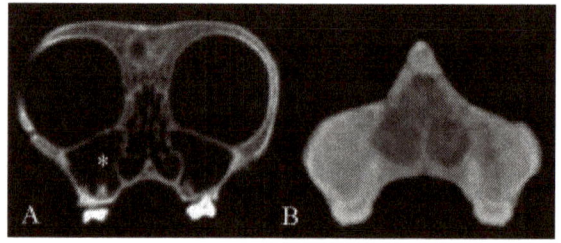

Figure 2 - Coronal CT scans (1mm) through M² of A) *Hylobates* (gibbon), and B) *Victoriapithecus*. In the gibbon, the maxillary sinus (*) can clearly be seen lateral of the nasal cavity proper, and separated from the nasal cavity by a wall of bone. In the fossil stem cercopithecoid, no maxillary sinus is present, and the same region consists entirely of spongy bone.

3D RECONSTRUCTION AND QUANTIFICATION

Results for studies of presence/absence characters can be achieved without reconstructing the specimens; visual inspection of the scans alone is necessary. This is not the case, however, if quantification and/or detailed shape information are required. Both the digital nature of CT data and the availability of commercial software packages for reconstruction/measurement have vastly improved the speed and accuracy with which detailed quantitative information can be obtained for characteristics previously unavailable for large scale study. This information can be used to test previous hypotheses based on qualitative assessments of internal cranial features, such as those of sinus volume.

Previous phylogenetic analyses of hominoid primates, or apes (including humans), have included the size of the maxillary sinus as a character supporting various arrangements of the extant and extinct members of the group (Andrews & Martin, 1987; Begun, 1992; Begun, et al., 1997; Harrison, 1987; Rae, 1997). In all, four separate scenarios have been advanced, advocating as few as zero and as many as two character state changes across ape evolution. Thus, a consensus on the pattern of evolution of cranial pneumatization in hominoids has failed to emerge from qualitative analyses, partially due to disagreement over the character state of individual taxa; for example, *Pongo* (the orang-utan) has been interpreted as having a moderately sized maxillary sinus, an expanded sinus, and a greatly expanded sinus (see Rae, 1999). One of the main barriers to such a consensus is that qualitative characters are difficult to interpret when confounding variables such as body size are present; extant hominoid taxa average from 5-125kg (Fleagle, 1999).

Again, new techniques of visualization and quantification can be used to test between competing hypotheses. In the case of ape sinus evolution, 3D reconstructions from CT scans can produce a quantitative evaluation of maxillary sinus volume which, when regressed against a body size indicator, provides an accurate, scaled assessment of the character state for the taxa concerned. The authors (Rae & Koppe, 2000b) performed such a test, using CT analysis of 44 adult dry crania representing all five extant hominoid genera. 3D virtual reconstructions of the crania were obtained using the ALLEGRO graphics workstation (ISG Technologies), and sinus volume was regressed against various measures (2D and 3D) of cranial size.

Contrary to most qualitative evaluations of hominoid maxillary sinus size, the quantitative data support no change across the group (Figure 3). Sinus volume is both highly and significantly correlated with cranial size, and scales isometrically with volumetric measures of the craniofacial skeleton (Rae & Koppe, 2000b).

In this case, 3D reconstruction allows the rejection of hypotheses of morphological change across a topology, and provides a more accurate, scaled character state delineation, both of which suggest that evolutionary trees constructed using maxillary sinus change in hominoids as supporting evidence must be critically re-evaluated. In turn, this may have a significant effect on the placement of fossil taxa, particularly where they have been allocated to clades based in part on the assumed homology of maxillary pneumatization (e.g., Begun, 1992).

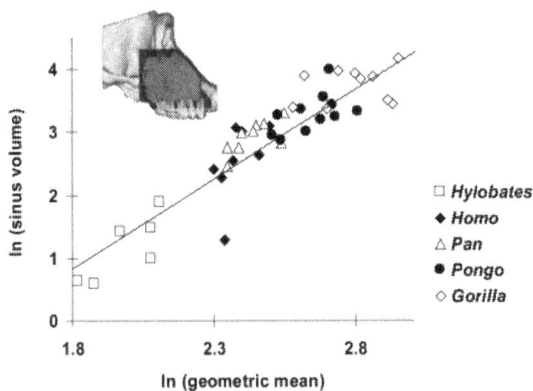

Figure 3 - 3D-CT-derived maxillary sinus volume regressed against size of the craniofacial skeleton ('geometric mean') in Hominoidea (apes, including humans). The 3D-CT reconstruction shown is an adult *Pan* (chimpanzee). Sinus size is significantly correlated with craniofacial size ($r = 0.89$, $p < 0.01$), and the slope of the reduced major axis regression line is not significantly different from isometry. The relationship suggests that no change in sinus volume has occurred in the evolution of extant hominoids. Adapted from Rae & Koppe (2000b).

GROWTH STUDIES

According to the functional matrix hypothesis (Moss & Young, 1960), the skull is thought of as a complex of different components with distinct functions. For each of these functions, such as chewing or respiration, particular functional cranial components have been identified. These components are morphological entities that consist of two parts: (1) the soft tissue that carries out a certain function, and (2) the skeletal unit that support and protect the functional matrix. It is important to consider that each of these skeletal units is characterized by a distinct growth pattern. Because it is held

that the differences in the growth velocity of particular parts of the body may produce differences in the morphology of different species (Tanner, 1992), the appreciation of both the individual growth pattern of each of these skeletal units as well as the interaction among the skeletal units is required.

It has been suggested previously (Ward & Brown, 1986) that variation in skull pneumatization among primates, and especially the development of the maxillary sinus, is a function of body size / skull size. In contrast to hominoids and numerous New World monkeys, however, some species of the genus *Macaca* possess pneumatic spaces that are quite small for primates of their body / skull size (Koppe, et al., 1999c). One way to elucidate the differences in the skull pneumatization patterns between macaques and other groups of primates is to test whether the morphology as seen in adults is the result of distinct growth patterns. Therefore, we have used CT data of two cross-sectional series of skulls to compare the relative growth of the maxillary sinus of the Japanese macaque (*Macaca fuscata*) with that of the orang-utan (*Pongo pygmaeus*). Since the maxillary sinus develops via an epithelial recess from the nasal mucus membrane, we also tested whether the relationship between the maxillary sinus and the growing nasal cavity differs between these two species.

Maxillary sinus – The comparison of the two mixed-sex samples of dry crania (*M. fuscata*, $n = 30$; *P. pygmaeus*, $n = 40$) reveals that within a given species both male and female maxillary sinuses tend to grow according to a common growth pattern. Although both species are characterized by a distinct sexual size dimorphism in numerous parts of the skull (Masterson & Leutenegger, 1992; Mouri, 1994), sexual dimorphism in maxillary sinus volume is only observed in the orang-utan.

The comparison of the relative growth of the maxillary sinus of *M. fuscata* with that of *P. pygmaeus*, using basicranial length as a surrogate of skull size, indicates that the maxillary sinus of the orang-utan enlarged postnatally faster and over a longer period than that of the Japanese macaque (Figure 4a). This suggests that differences in the postnatal growth pattern may be responsible in part for the variation in the degree of skull pneumatization among primate species.

Nasal cavity - In primates, the nasal cavity serves mainly as an integrative part of the respiratory tract, and it has been suggested that the growth of the nasal cavity in humans is closely related to the development of the paranasal sinuses. Because little information is available about this association in nonhuman primates, we investigated the implication of the nasal cavity on the growth of the maxillary sinus in catarrhine primates. The CT scans of the skulls of the two cross-sectional series described above served were also used to calculate nasal cavity volume for *M. fuscata* and *P. pygmaeus*. Reduced major axis analysis was applied to describe the growth pattern of the nasal cavity volume and to investigate the relationship between the nasal cavity volume and maxillary sinus volume. In contrast to the differences in the degree of sexual dimorphism in sinus size *between* the species (see above), sexual dimorphism in nasal cavity volume is seen in both the orang-utan and the macaque.

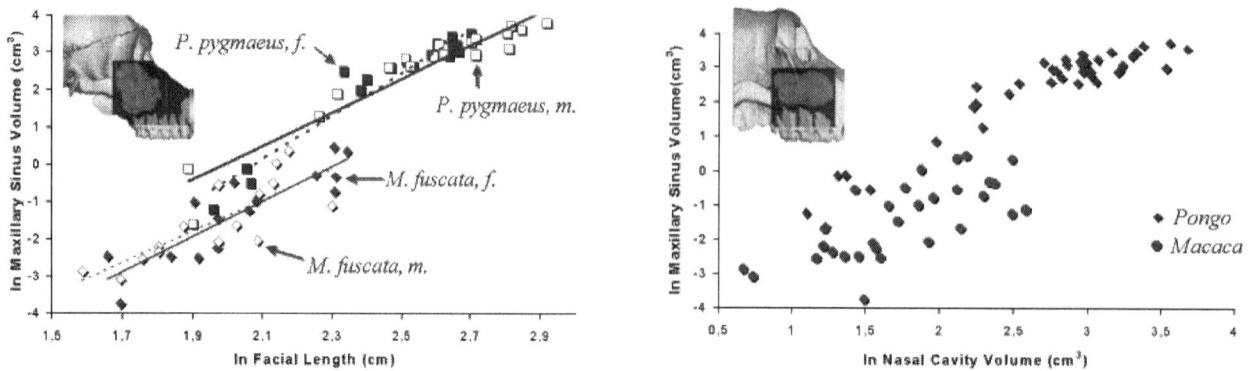

Figure 4 - A) Growth of the maxillary sinus in *Pongo pygmaeus* and *Macaca fuscata*. The trajectories of the growth curves are similar, although the orang-utan sinus grows at a faster rate and for longer than that of the macaque. The 3D-CT reconstruction pictured is an adult *Pongo*.
B) Relationship between growth in the nasal cavity and maxillary sinus in *P. pygmaeus* and *M. fuscata*. The regularity of this relationship, along with the differences seen in the growth of the sinus (A), suggests a common underlying mechanism. The 3D-CT reconstruction pictured is an adult *Macaca*.

In both *P. pygmaeus* and *M. fuscata*, maxillary sinus volume enlarged with a steeper slope than the nasal cavity volume. In contrast to maxillary sinus volume, however, the interspecies comparison (sexes were pooled) revealed no significant differences in the regression slopes of the nasal cavity volume. Although the growth rate of the maxillary sinus volume of *M. fuscata* was significantly less than that of the orang-utan, the two-tailed t-test indicated no significant differences for the relationship between the nasal cavity volume and the maxillary sinus volume (Figure 4b). The similarity in the growth rates of the maxillary sinus in *P. pygmaeus* and *M. fuscata* relative to the growth of the nasal cavity, and the disassociation of the growth of the maxillary sinus volume from that of the nasal cavity, suggests that the underlying mechanism for the development of paranasal pneumatization is similar among anthropoid primates. Notwithstanding that the maxillary sinus is closely associated to the nasal cavity, variation in size and shape of the maxillary sinus among primates is not related exclusively to respiratory function. Studies of different local populations of the Japanese macaque (Koppe, et al., 1999b) as well as studies in humans (Shea, 1977) suggest that epigenetic factors influence the final size and shape of the pneumatic spaces.

DISCUSSION

Because of the difficulties of visualizing and measuring internal structures, pneumatization of the mammalian cranium has remained a relatively unexplored area, despite the fact that sinus spaces were first described three and half centuries ago (Koppe, et al., 1999a). In the past, workers tended to rely on data derived from destructive methods, such as dissection, which limited the size of the data sets that could be obtained. Small sample size compromised even some standard radiographic evaluations of primate sinuses (Lund, 1988), from which misleading terms such as "lateral recess" (Rae & Koppe, 2000a) were derived. Even so, the efforts of Paulli (1900) and later Cave & Haines (1940) were considerable, though still not enough to generate the

kind of momentum necessary for a substantial body of research to emerge.

As the studies outlined above make evident, the advent of 3D-CT marks a significant turning point in the study of the mammalian cranium. This has become apparent even in areas other than paranasal pneumatization, such as the study of the bony labyrinth of the ear (Spoor & Zonneveld, 1998). The potential for this method to address long-standing questions, such as the homology of the frontal sinus in anthropoid primates (Ward & Pilbeam, 1983) and the effect of environmental factors on sinus size (Shea, 1977) leaves little doubt that 3D-CT will continue to stimulate new study. Inevitably, our understanding of both the functional (Koppe, et al., 1999c) and phylogenetic (Rae, 1999) aspects of the heretofore hidden internal constitution of the cranium will increase as more researchers (e.g., Márquez, et al., 1997; Rossie, 2000) begin to apply these techniques and procedures.

The degree to which any new method is accepted in science is directly proportional to the number of hypotheses that can be tested and generated from its use. Thus, the application of novel approaches will inevitably result in the critical re-evaluation of previous work, and the development of new avenues of exploration. 3D-CT has already contributed a great deal to the scientific study of the mammalian skull, both analytically and clinically (e.g., Uchida, et al., 1998), and should continue to provide insights towards our understanding of the evolution of cranial pneumatization.

Acknowledgements

The authors express their sincere regret for being unable to attend the conference, but thank the convenors, Bertrand Mafart and Hervé Delingette for the invitation to participate in both the symposium and the present volume. Thanks are also due to J. Balczun (Essen) and R. Reike (Crimmitschau) for their help in CT analysis. Work reported here was sponsored by grants to TCR from the Leakey Foundation (USA) and the Univ. of Durham, and to TK from the Primate Research Institute of Kyoto University (Japan) and the Univ. of Greifswald.

Authors' addresses:

Todd C. RAE - Evolutionary Anthropology Research Group, Dept. of Anthropology, Univ. of Durham, 43 Old Elvet, Durham DH1 3HN, UK, and Dept. of Mammalogy, American Museum of Natural History, Central Park West at 79th St., New York, New York, 10023, USA

Thomas KOPPE - Institut für Anatomie, Ernst-Moritz-Arndt-Universität Greifswald, Friedrich-Loeffler-Str. 23 c, D-17487 Greifswald, Germany

BIBLIOGRAPHY

ANDREWS, P. & MARTIN, L., 1987, Cladistic relationships of extant and fossil hominoids. *Journal of Human Evolution* 16, p. 101-118.

AX, P., 1985, Stem species and the stem lineage concept. *Cladistics* 1, p. 279-287.

BEGUN, D., 1992, Miocene fossil hominids and the chimp-human clade. *Science* 257, p. 1929-1933.

BEGUN, D., WARD, C. & ROSE, M., 1997, Events in hominoid evolution. In *Function, phylogeny, and fossils: Miocene hominoid evolution and adaptations*, edited by D. Begun, C. Ward & M. Rose. New York: Plenum, p. 389-415.

BENEFIT, B. & MCCROSSIN, M., 1991, Ancestral facial morphology of Old World higher primates. *Procedings of the National Academy of Sciences USA* 88, p. 5267-5271.

BENEFIT, B., 1993, The permanent dentition and phylogenetic position of *Victoriapithecus* from Maboko Island, Kenya. *Journal of Human Evolution* 25, p. 83-172.

BENEFIT, B. & MCCROSSIN, M., 1993, Facial anatomy of *Victoriapithecus* and its relevance to the ancestral cranial morphology of Old World monkeys and apes. *American Journal of Physical Anthropology* 92, p. 329-370.

BENEFIT, B. & MCCROSSIN, M., 1997, Earliest known Old World monkey skull. *Nature* 388, p. 368-371.

BLANEY, S., 1986, An allometric study of the frontal sinus in *Gorilla, Pan,* and *Pongo. Folia Primatologica* 47, p. 81-96.

BLANEY, S., 2000, Scaling properties of the frontal sinus in the african great apes - a clue to the role of the human paranasal sinuses. *Revue de Laryngologie -Otologie - Rhinologie* 121, p. 99-102.

CAVE, A. & HAINES, R., 1940, The paranasal sinuses of the anthropoid apes. *Journal of Anatomy* 74, p. 493-523.

HARRISON, T., 1987, The phylogenetic relationships of the early catarrhine primates: a review of the current evidence. *Journal of Human Evolution* 16, p. 41-80.

KOPPE, T. & NAGAI, H., 1995, On the morphology of the maxillary sinus floor in Old World monkeys - a study based on three-dimensional reconstructions of CT scans. In *Proceedings of the 10th international symposium on dental morphology*, edited by R. Radlanski & R. Renz. Berlin: C. & M. Brünne GbR., p. 423-427.

KOPPE, T., INOUE, Y., HIRAKI, Y. & NAGAI, H., 1996, The pneumatization of the facial skeleton in the Japanese macaque (*Macaca fuscata*) - a study based on computerized three-dimensional reconstructions. *Anthropological Science* 104, p. 31-41.

KOPPE, T. & NAGAI, H., 1999, Quantitative analysis of the maxillary sinus in catarrhine primates. In *The paranasal sinuses of higher primates: development, function and evolution*, edited by T. Koppe, H. Nagai & K. Alt. Berlin: Quintessence, p. 121-149.

KOPPE, T., NAGAI, H. & ALT, K., 1999a, Introduction. In *The paranasal sinuses of higher primates: development, function and evolution*, edited by T. Koppe, H. Nagai & K. Alt. Berlin: Quintessence, p. 15-20.

KOPPE, T. & OHKAWA, Y., 1999, Pneumatization of the facial skeleton in catarrhine primates. In *The paranasal sinuses of higher primates: development, function and evolution*, edited by T. KOPPE, H. NAGAI & K. Alt. Berlin: Quintessence, p. 77-119.

KOPPE, T., RAE, T. & MARQUEZ, S., 1999b, Determinants of the variation of the maxillary sinus size in Japanese macaques (abstract). *American Journal of Physical Anthropology* Suppl. 28, p. 173.

KOPPE, T., RAE, T. & SWINDLER, D., 1999c, Influence of craniofacial morphology on primate paranasal pneumatization. *Annals of Anatomy* 181, p. 77-80.

LUND, V., 1988, The maxillary sinus in the higher primates. *Acta Otolaryngologica* 105, p. 163-171.

MÁRQUEZ, S., GANNON, P., REIDENBERG, J., DELSON, E. & LAITMAN, J., 1997, Relationship between maxillary sinus volume and craniofacial linear measurements in *Macaca fascicularis* and *Macaca mulatta* (abstract). *American Journal of Physical Anthropology* Suppl. 24, p. 161.

MASTERSON, T. & LEUTENEGGER, W., 1992, Ontogenetic patterns of sexual dimorphism in the cranium of Bornean orang-utans (*Pongo pygmaeus pygmaeus*). *Journal of Human Evolution* 23, p. 3-26.

MOSS, M. & YOUNG, R., 1960, A functional approach to craniology. *American Journal of Physical Anthropology* 18, p. 281-292.

MOURI, T., 1994, Postnatal growth and sexual dimorphism in the skull of the Japanese macaque (*Macaca fuscata*). *Anthropological Science* 102 (Suppl.), p. 43- 56.

NOVACEK, M., 1993, Patterns of diversity in the mammalian skull. In *The skull, vol. 2: patterns of structural and systematic diversity*, edited by J. Hanken & B. Hall. Chicago: Univ. of Chicago Press, p. 438-545.

PAULLI, S., 1900, Über die Pneumaticität des Schädels bei den Säugethieren. III. Über die Morphologie des Siebbeins und Pneumaticität bei den Inxectivoren, Hyracoideen, Chiropteren, Canivoren, Pinnipedien, Edentaten, Rodentiern, Prosimien und Primaten. *Morphol. Jb.* 28, p. 483-564.

RAE, T., 1997, The early evolution of the hominoid face. In *Function, phylogeny, and fossils: Miocene hominoid evolution and adaptations*, edited by D. Begun, C. Ward & M. Rose. New York: Plenum, p. 59-77.

RAE, T., 1999, The maxillary sinus in primate paleontology and systematics. In *The paranasal sinuses of higher primates: development, function and evolution*, edited by T. Koppe, H. Nagai & K. Alt. Berlin: Quintessence, p. 177-189.

RAE, T. & KOPPE, T., 2000a, Definition of the "lateral recess" and cranial pneumatization in the Catarrhini (abstract). *American Journal of Physical Anthropology* Suppl. 30, p. 257.

RAE, T. & KOPPE, T., 2000b, Isometric scaling of maxillary sinus volume in hominoids. *Journal of Human Evolution* 38, p. 411-423.

RAE, T., KOPPE, T., SPOOR, F., BENEFIT, B. & MCCROSSIN, M., in press, Ancestral loss of the maxillary sinus in Old World monkeys and independent acquisition in *Macaca. American Journal of Physical Anthropology*

ROSSIE, J., 2000, Platyrrhine paranasal sinus patterns: a preliminary investigation (abstract). *American Journal of Physical Anthropology* Suppl. 30, p. 265.

SHEA, B., 1977, Eskimo craniofacial morphology, cold stress and the maxillary sinus. *American Journal of Physical Anthropology* 47, p. 289-300.

SPOOR, F. & ZONNEVELD, F., 1998, Comparative review of the human bony labyrinth. *Yearbook of Physical Anthropology* 41, p. 211-251.

SPOOR, F. & ZONNEVELD, F., 1999, Computed tomography-based three-dimensional imaging of hominid fossils: features of the Broken Hill 1, Wadjak 1, and SK 47 crania. In *The paranasal sinuses of higher primates: development, function and evolution*, edited by T. Koppe, H. Nagai & K. Alt. Berlin: Quintessence, p. 207-226.

TANNER, J., 1992, Human growth and development. In *The Cambridge encyclopedia of human evolution*, edited by S. Jones, R. Martin & D. Pilbeam. Cambridge: Cambridge Univ Press, p. 98-105.

UCHIDA, Y., GOTO, M., KATSUKI, T. & SOEJIMA, Y., 1998, Measurement of maxillary sinus volume using computerized tomographic images. *International Journal of Oral & Maxillofacial Implants* 13, p. 811-818.

VLČEK, E., 1967, Die Sinus frontales bei europäischen Neandertalern. *Anthropologischer Anzeiger* 30, p. 166-189.

WARD, S. & PILBEAM, D., 1983, Maxillofacial morphology of Miocene hominoids from Africa and Indo-Pakistan. In *New interpretations of ape and human ancestry*, edited by R. Ciochon & R. Corruccini. New York: Plenum, p. 211-238.

WARD, S. & BROWN, B., 1986, The facial skeleton of *Sivapithecus indicus*. In *Comparative primate biology, vol. 1: systematics, evolution, and anatomy*, edited by D. Swindler & J. Erwin. New York: Alan R. Liss, p. 413-452.

WITMER, L., 1997, The evolution of the antorbital cavity of archosaurs: a study in soft-tissue reconstruction in the fossil record with an analysis of the function of pneumaticity. *Journal of Vertebrate Paleontology* 17 (Suppl. to No. 1), p. 1-73.

WITMER, L., 1999, The phylogenetic history of paranasal sinuses. In *The paranasal sinuses of higher primates: development, function and evolution*, edited by T. Koppe, H. Nagai & K. Alt. Berlin: Quintessence, p. 21-34.

THE "VIRTUAL" ENDOCAST OF SACCOPASTORE 1. GENERAL MORPHOLOGY AND PRELIMINARY COMPARISONS BY GEOMETRIC MORPHOMETRICS

Emiliano BRUNER, Giorgio MANZI & Pietro PASSARELLO

Résumé: L'endocrâne du pré-Neandertal Saccopastore 1, obtenu par scannerisation CT de l'exemplaire original puis reconstruction virtuelle des volumes internes, est comparé avec d'autres hommes fossiles du Pléistocène Moyen et Supérieur. Les caractéristiques morphologiques générales sont analysées à partir d'une sélection de valeurs métriques par morphométrie géométrique. L'endocrâne de Saccopastore 1 présente des traits primitifs: platycéphalie, rostrum encéphalique, position de la largeur maximum, configuration du réseau méningé et des circonvolutions cérébrales, etc. En même temps, ces traits sont associés à des aspects de type Néandertalien, particulièrement quand on considère les largeurs, les structures occipito-cérébelleuses, la latéralisation et les asymétries. A partir d'une approche basée sur la morphométrie géométrique, la variabilité morphologique endocrânienne à l'intérieur du genre Homo apparaît être fortement liée à la taille, et, pour les faces latérales, à l'expansion des zones pariétales. Dans ce cadre général, le caractère plésiomorphe de l'endocrâne de Saccopastore 1 est resitué dans une perspective allométrique et l'hypothèse d'une pause dans l'évolution du cerveau (en dehors de son volume) pendant une grande partie de l'histoire naturelle du genre Homo est discutée brièvement.

Abstract: The endocast of the early Neandertal specimen Saccopastore 1 – obtained by CT-scanning of the original specimen and virtual reconstruction of the internal volumes – is compared with other fossil specimens ranging between Middle and Late Pleistocene. Features of general morphology are reported, together with a selection of metrical values and preliminary results of an ongoing geometric morphometric analysis. The endocast of Saccopastore 1 shows some primitive traits: platicephaly, encephalic rostrum, maximum breadth location, patterns of both meningeal system and gyrification system, etc. At the same time, these features are blended with derived Neandertal traits, particularly when width dimensions, occipito-cerebellar structures, as well as lateralisation and asymmetries are considered. From the multivariate approach based on geometric morphometrics, variability in endocranial vault morphology within the genus Homo appears to be size-related. In this general framework, the plesiomorphic appearance of the endocast of Saccopastore 1 is evaluated from an allometric perspective, and the case for stasis in brain evolution (but for its size) during great part of the natural history of the genus Homo is also briefly discussed.

INTRODUCTION

The fossil cranium known as Saccopastore 1 (Scp.1) was recovered in 1929 in the homonymous area now within Rome (Sergi, 1929, 1944), included in a gravel/sand deposit referred to the beginning of the Late Pleistocene, that is to about 100-120 ka (for a review of the literature, see Manzi & Passarello, 1991). It is well known that Scp.1, commonly considered a female specimen, exhibits a morphological pattern that appears consistent with its chronology, according to the so-called 'accretion model' for human evolution in Europe (Hublin, 1998). Clearly derived – i.e., Neandertal – traits are in fact blended with more archaic and less derived features (e.g., Condemi, 1992), including small cranial capacity (1174 ml was the best estimate obtained by Sergi, 1944) and marked platicephaly; this "mosaic" pattern is shared with Middle Pleistocene samples along the Neandertal lineage (compare Arsuaga et al., 1997). The fossil shows an extremely high level of mineralisation, and the endocranial cavity is partially filled by the geological matrix of inclusion.

The cranium has been recently CT scanned (Manzi et al., 2001) to allow, among other aims, the analysis of internal structures. In this framework, the endocast was reconstructed by computed assisted imaging techniques (Conroy & Vannier 1984; Zollikofer et al. 1998; Recheis at al. 1999; Spoor et al. 2000). In this paper, we report some notes on general morphology of the endocast of Scp.1 and a selection of metrical values, together with preliminary results obtained in comparing other fossil endocasts by geometric morphometrics. In perspective, this work is aimed to characterise the architecture of the Scp.1 endocranial morphology – and the decomposition of its shape – in the scenario of the evolution of the genus *Homo*. Clearly, the statistical power of this preliminary analysis must be increased by a larger sample, that will include other human *taxa*, samples, and specimens (Bruner, n.d.). Temporal areas must be added to the analysis, and a special attention should be paid to the cerebellar lobes (functionally linked to the parietal structures). Moreover, the approached based on geometric morphometrics will be extended to the superior view (which accounts for cerebral widths that, in turn, better characterise the Neandertal morphology) and/or to the three-dimensional space (for a description of the whole system).

MATERIALS AND METHODS

Scp.1 was scanned using a Tomoscan AUEP (Phylips), with sequential and contiguous 1 mm scans, 75 mA and 140 kV, and 0.49 mm pixel size (Manzi et al. 2001). The skull has been scanned by transverse planes, according to the Frankfurt horizontal. Data were exported as DICOM files. The high level of fossilisation, the stone matrix inclusions and the large depth of the layers caused marked streak artefacts and diffuse noise. Therefore, besides increasing the beam power (mA), a filter was necessarily used to clean the signal. Data have been analysed using MIMICS 7.0 package (Materialise). Volumes have been segmented by thresholding the different CT numbers and following, in general, the "half maximum height" technique (Spoor et al., 1993).

Spatial coordinates of landmarks were analysed by Generalised Procrustes Analysis and Thin-Plate-Spline interpolation (Rohlf & Bookstein 1990; Bookstein 1991; Marcus et al. 1993; Rohlf 1993; Rohlf & Marcus 1993; Lynch et al. 1996; Marcus et al. 1996), using TPS softwares (Rohlf 1997, 1998a, 1998b, 2000a, 2000b) and the APS package (Penin, 2000). The landmarks were chosen in order to optimise the number of specimens available, and referred mainly to the vault profile in lateral view (Fig. 1). Two dimensional coordinates were collected without parallax distorsion from homologous landmarks (prerolandic and transverse sulci on midsagittal profile), points of maximum curvature (frontal and occipital poles, Broca area), and orthogonal projections from chords between the previous ones (frontal and parietal chords). The Scp.1 virtual endocast was compared with physical endocasts available at the Museum of Anthropology "G. Sergi" and the Istituto Italiano di Paleontologia Umana, in Rome. The comparative sample is composed by Trinil 2, Salé, Zhoukoudian III and XII, Arago (reconstruction), Neanderthal 1, La Chapelle-aux-Saints, and Guattari endocasts. Coordinates on comparative specimens were collected using a dioptograph and the tpsDig software (Rohlf, 1998c). The analysis based on the unweighted pair group method using arithmetic averages (UPGMA) was computed with PHYLIP 3.57c package (Felsenstein, 1989), and phenograms were elaborated by TREEVIEW (Page, 1996).

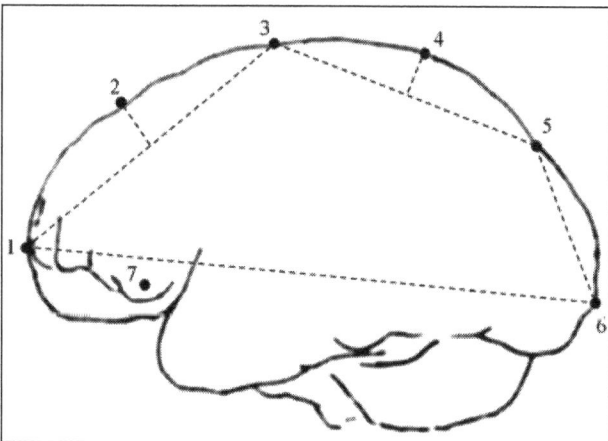

Fig. 1 - Landmarks sampled in lateral view. 1) most anterior point of the frontal pole; 2) orthogonal projection at the mid chord 1-3; 3) prefrontal sulcus; 4) orthogonal projection at the mid chord 3-5; 5) perpendicular sulcus; 6) most posterior point of the occipital lobe; 7) Broca area, at the boss between *pars opercularis* and *pars triangularis*.

RESULTS

General Morphology

Scp.1 is a heavily fossilised specimen (average HU = 3012 ± 622); in addition, the geological matrix fills all the

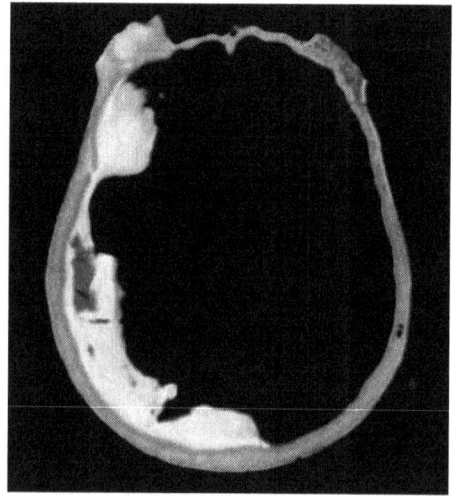

Fig. 2 - Transversal slice of Saccopastore 1 showing the stone matrix included in the endocranial cavity.

internal cavities. Even the endocranial surface is covered by sediment in various areas of the base and both the left and posterior walls of the braincase (Fig. 2), and particularly: the proximal areas of the right 3rd frontal circumvolution, the right temporal lobe, the cerebellar poles, and the occipital poles. Nevertheless, the cerebral surfaces are well preserved (especially those covered by the geological matrix), and the entire structure can be successfully reproduced (Fig. 3) except for some basal areas, where the sediment appears extremely interconnected with the fossil matrix.

The endocranial cavity is clearly asymmetric, with a strong dominance of the left hemisphere. The left frontal lobe is in fact slightly longer and broader, and the left parietal and occipital areas are definitely larger than the right ones in both the transvesal and coronal planes. Moreover, the maximum breadth is localised on the right at about mid-height of the hemisphere (temporo-parietal level), while on the left side it is placed forward and in a definitely lower position (base of the 3rd temporal circumvolution).

The frontal lobes show a marked encephalic rostrum, resembling the stage 1 described by Grimaud-Hervé (1997). The prefrontal circumvolutions are well defined, being more expressed on the left side, and the Broca area is clearly developed on the same hemisphere. The orbital circumvolutions are not identifiable, but they seem localised roughly on the orbital roofs, and not behind them. Cerebral impressions of the temporo-parietal areas are not easily discernible. The boss corresponding to the area of the angular and supramarginal gyri can be recognised on the left side, being the right one more rounded and smoother. Occipital lobes are only moderately developed, without any strong angulation with the parietal area. Cerebellar lobes, rather globular in shape, are placed anteriorly to the occipital poles, standing for the most part below the parietal area, reciprocally in contact and close to the mid-sagittal plane.

As far as vascular relieves and sinuses are concerned, a marked wrinkle represents the anterior branch of the meningeal system on the left hemisphere, and it could be interpreted as a spheno-parietal sinus, being a smooth, unbranched and enlarging track

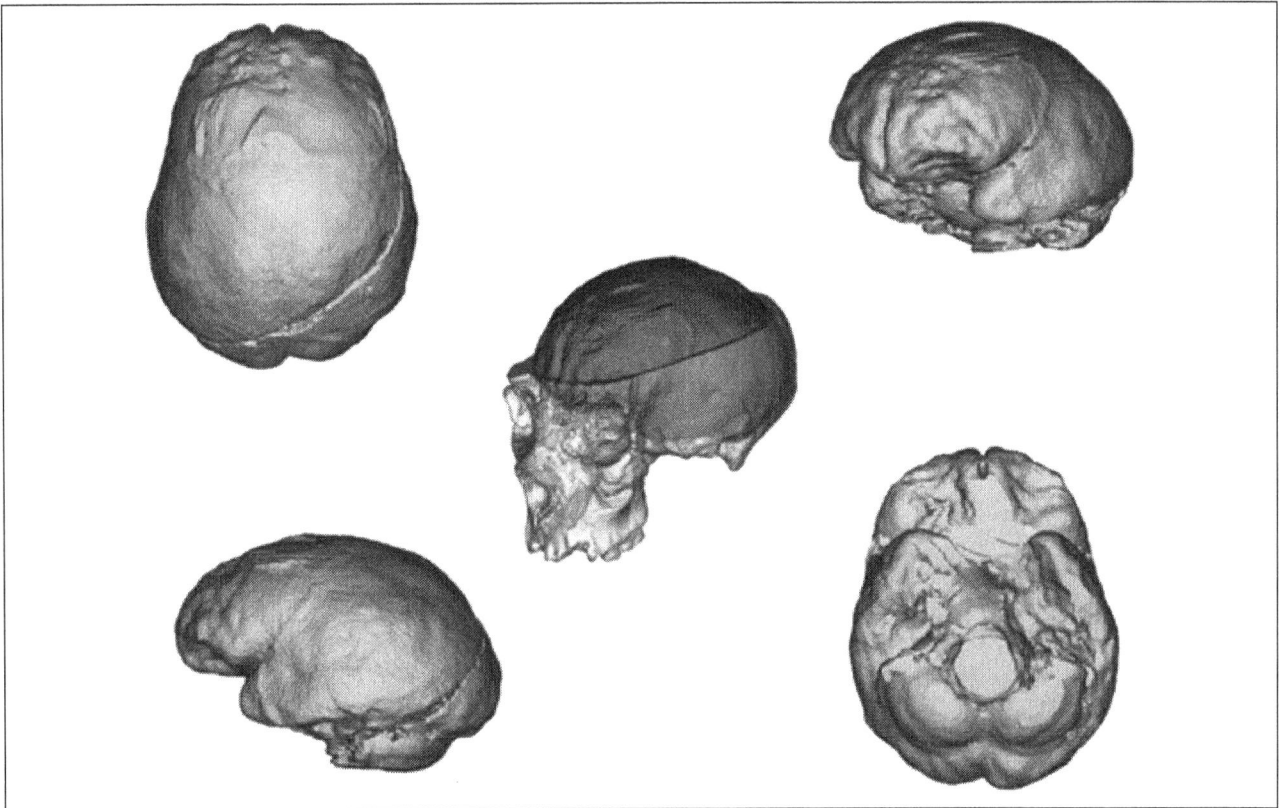

Fig. 3 - Virtual replicas of the Saccopastore 1 endocast.

that approaches the coronal suture; however, its superior end is not fully detectable, reaching the area of a large hole opened in the braincase. In these conditions, the occurrence of the Breschet sinus cannot be unquestionably assessed (D. Grimaud-Hervé, pers. comm.), but rather possible. On the right side, the anterior branch of the meningeal system is hardly noticeable, while the posterior one is more marked. The obelic (middle) ramus splits from the posterior one, branching at least once, and representing the most visible and large impression of the right meningeal system. The lambdatic wrinkle is marked but not branched. Whether this reconstruction of the fragmentary evidence observed on the right hemisphere is correct, Saccopastore 1 shows a configuration of Adachi type II (compare Falk, 1993). Anastomoses are not detected.

The superior sagittal sinus is clearly visible only in the interparietal region, showing a diameter of about 8 mm. At the *torcular herophili*, the sagittal sinus runs clearly into the right, well developed, transverse-sigmoid system. Conversely, the left transverse sinus is hardly detectable, and the area of the respective sigmoid sinus is not well preserved. No occipito-marginal system is noticed.

COMPARISONS: METRICAL DATA AND GEOMETRIC MORPHOMETRICS

Scp.1 has a rather small endocast, with an average hemispheric maximum length of 161 mm, maximum breadth of 131 mm, width at the Broca's cap of 102 mm,

and height (endobasion-endovertex) equal to 103 mm. In Figure 4, metrical values of Scp.1 are plotted with average values of fossil samples ranging between Middle and Late Pleistocene (data from Grimaud-Hervé, 1997). Scp.1 shows both hemispheric length and cerebral height clearly bracketed within the range of variability of specimens referred to *Homo erectus* (Sangiran and Zhoukoudian). Conversely, cerebral widths are close to values observed in the Ngandong sample, standing in an intermediate position between the European Middle Pleistocene sample, from one side, and Neandertals and modern humans, from the other. Anyway, when reported relatively to the exocranial dimensions, length and width measurements show values that are close to those expressed by Wurmian Neandertals.

In Fig. 5a, the specimens submitted to geometric morphometric analysis are plotted in the plane described by the first two principal components. As a result of the shape decomposition, these two principal components explain together 80.2% of the total variance. It should be noted in the plot the position of Scp.1, intermediate within the *Homo erectus* cluster, with respect to typical Neandertals such as Guattari, La Chapelle, and the Feldhofer calotte.

The PC1 (44.8% of variance) involves flattening of the posterior (parieto-occipital) areas (Fig. 5b), while PC2 (35.4%) relates to the relative reduction of the prefrontal region (Fig.5c). A multivariate regression of shape versus size (represented by the centroid size of the mid-sagittal contour) shows a significant and marked correlation ($R^2 = 0.79$) between absolute dimensions and these two PCs, almost equally involved in a size-related effect. A Partial

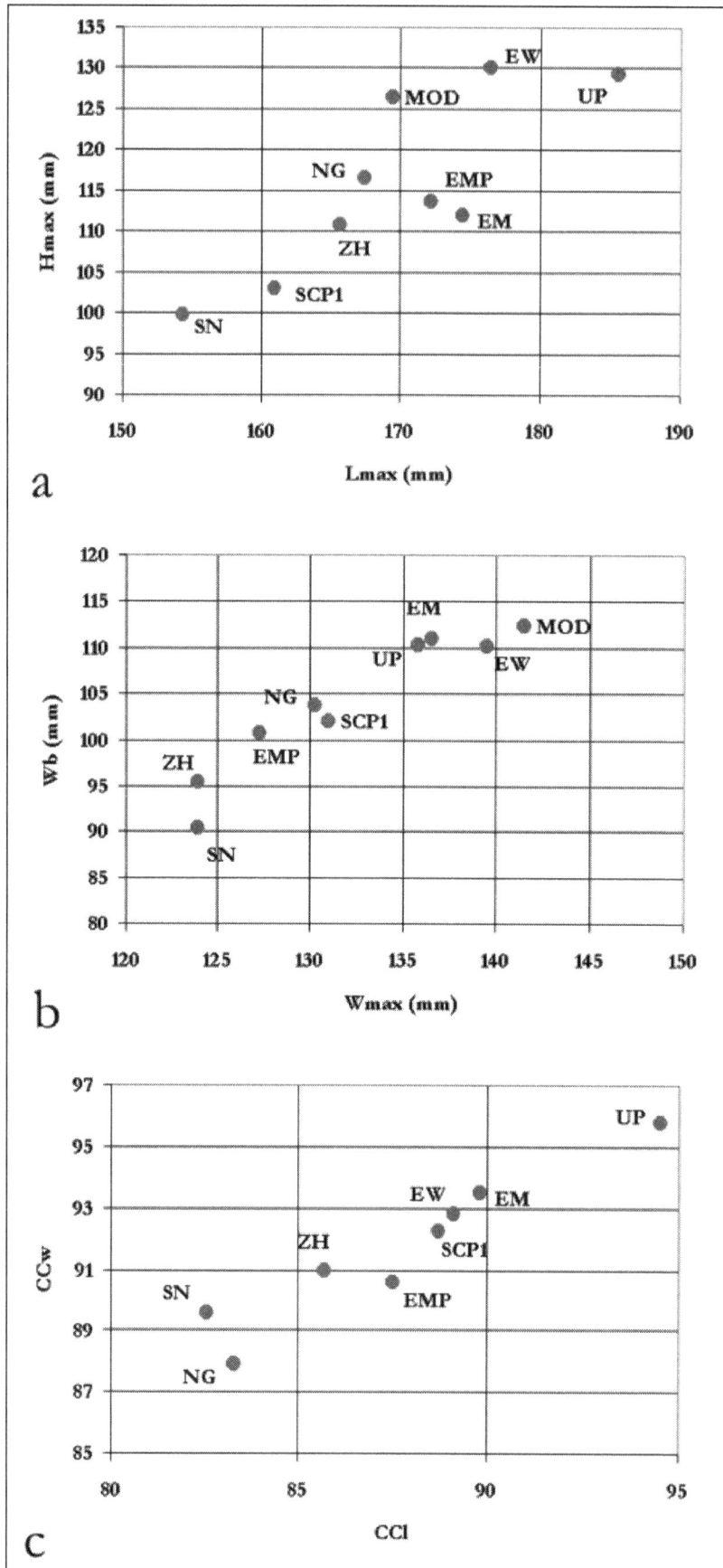

Fig. 4 - Comparison between endocranial dimensions of Saccopastore 1 and average values from other fossil samples (comparative samples and data from Grimaud-Hervé, 1997): a) average hemispheric length (Lmax) vs Endobasion-endovetrex height (Hmax); b) maximum width (Wmax) vs frontal width at the Broca's cap (Wb); c) cranio-encephalic length index (CCl) vs cranio-encephalic width index (CCw). *Legend* (fossil samples) – SCP1: Saccopastore 1; SN: Sangiran; ZH: Zhoukoudian; NG: Ngandong; EMP: European Middle Pleistocene; EW: European Würmian Neandertals; EM: early modern humans; UP: Upper Paleolithic; MOD: modern humans.

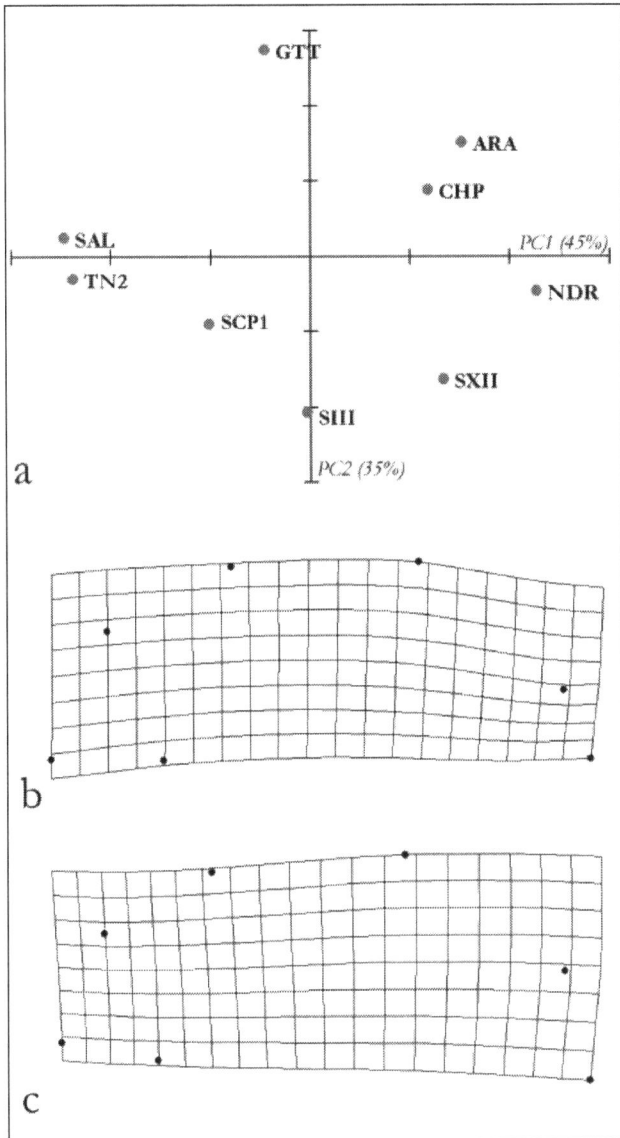

a

b

c

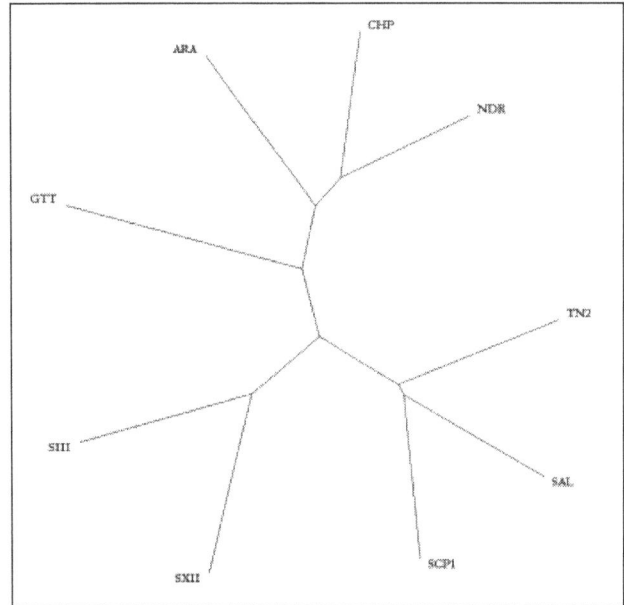

Fig. 6 - Unrooted tree resulted from an unweighted pair group method using arithmetic averages (UPGMA) analysis based on Procrustes distances; labels as in Figure 5.

Fig. 5 - Plotting of the sample in the space of the first two principal components (a). Distortion grids showing the spline along the PC1 (b) and PC2 (c) are also reported. The basic configuration of landmarks is reported in Figure 1. *Legend* (fossil specimens) – SCP1: Saccopastore 1; SAL: Salè; TN2: Trinil 2; SIII: Zhoukoudian (Sinanthropus III); SXII: Zhoukoudian (Sinanthropus XII); ARA: Arago (reconstruction); NDR: Neandertal 1 (Feldhofer); CHP: La Chapelle-aux-Saints; GTT: Guattari 1.

Least Square regression of size on the entire set of shape parameters shows the same degree of correlation and globally discriminates the shape of small endocasts against larger ones.

An UPGMA procedure was then applied to the Procrustes distances obtained from the geometric morphometric analysis reported above. In the resulting tree (Fig. 6), Neandertal specimens describe a cluster that also includes the Arago reconstruction, while Scp.1 relates to *Homo erectus* variability and is closer to smaller specimens (such as Trinil and Salé). In a second analysis (not reported here) the confluence of sinuses (*torcular herophili*) was added to the landmarks: the same results were found, although Scp.1 moved closer to specimens from Zhoukoudian.

DISCUSSION

It should be prudent when dealing with endocast morphologies based on reconstructions from CT-data (Zollikofer & Ponce de Leon, 2000). Yet, the virtual replica of Scp.1 appears complete and clear enough to allow a general description of the endocranial features, including size and both details and architecture of the shape. We can conclude that Scp.1 represents a useful test for this purpose, particularly when it is considered that it represents a specimen clearly within the Neandertal lineage, but with a brain scaled to the size of more archaic representatives of the genus *Homo*. As a matter of fact, the endocast of Scp.1 shows a mixture of features connecting the morphological patterns respectively observed in Middle and Late Pleistocene specimens.

For instance, although the shape of the frontal lobes seems to be of lesser importance during human evolution (Bookstein et al. 1999), Scp.1 shows a marked encephalic rostrum, that is considered a primitive feature shared within the *Homo erectus* hypodigm (Grimaud-Hervé, 1997). At the same time, diversely from what is reported for some large Middle Pleistocene specimens such as Petralona and Kabwe (Seidler et al., 1997), the orbital lobes in Scp.1 do not stand behind the frontal roof, but in a rather advanced position. Moreover, the frontal circumvolutions are well developed, including a large Broca area on the left side, and the ratio between frontal and parietal widths are fairly large when compared with Asian samples from Sangiran or Ngandong. Therefore – considering the common occurrence of strong platicephaly in Scp.1 as well as in specimens referred to *Homo erectus* – the expression of an encephalic rostrum could be viewed in relationship with the small endocranial heights, despite Scp.1 shows more derived traits in the frontal areas.

The occipital lobes are not stressed backwards as in the Asian clade, being rather continuous with the parietal outline and in close connection with the temporal lobes. The cerebellar lobes are globular and contiguous, while in *Homo erectus* they are more elliptical and detached (Grimaud-Hervé, 1997). This feature – already present in the Cranium 5 from Sima the los Huesos (EB, pers. obs.) – links Scp.1 to more derived configurations (Neandertals and modern humans), and could be in relationship with the longitudinal organisation of the whole system. During human evolution, all the posterior districts of the brain seem to have rotated gradually under the parieto-temporal areas. In Scp.1, as generally observed among Neandertals and (most of all) in modern humans, the cerebellar lobes lie almost entirely under the parietal ones, while in the Sangiran sample these are more posteriorly protruded (halfway between the parietal and the occipital areas); in more encephalized *Homo erectus* specimens, such as those from Zhoukoudian, the position is rather intermediate between these two extremes. Following such a process, along the general trend of human evolution, the cerebellar poles approach medially each other, changing the conformation of the entire cerebellum from a more longitudinal orientation to a little bit more transversal one.

Petalias are well developed in Scp.1. The global left dominance does not correspond to the more frequent right-frontal / left-occipital pattern (Holloway & de la Coste-Lareymondie, 1982), but it is nevertheless well expressed. The occurrence of cerebral asymmetries is another trend in human evolution – expression degree of petalias similar to Scp.1 have been found, without a marked gyral pattern, in small brained specimens such as Salé (Holloway 1981a) or the Indonesian *Homo erectus* clade (Holloway 1980, 1981b), but at a visual inspection lateralisation seems somewhat more developed in Scp.1. Actually, the asymmetries at the 3rd frontal circumvolution and at the supramarginal and angular gyri indicate an advanced level of lateralisation, with interesting functional consequences (Bradshaw, 1988).

The meningeal system is rather simple, and it is apparently similar to Asian (Zhoukoudian) and European (Arago) Middle Pleistocene specimens (Saban, 1995; Grimaud-Hervé, 1997), sharing a probable dominance of the posterior ramus, and lacking a complex branching pattern or the occurrence of diffused anastomoses.

In sum, the morphological pattern that can be observed dealing with the so far unknown endocast of Scp.1 appears peculiar and interesting for human evolution – *crucial* in connecting more archaic and more derived phenotypes. Strictly comparable with that of Scp.1 – characterised by the occurrence of a primitive encephalic rostrum and a simple meningeal structure, associated to a fully modern gyrification system and well developed cerebellar lobes – we have the pattern already described in Saccopastore 2, the less complete early Neandertal from the same site (Sergi, 1948).

Coming to the still preliminary comparative approach reported here, it should be observed that the main diameters (total length, width and height) set the position of Scp.1 – again – as an "intermediate" specimen between other fossil groups.

The endocast shows extreme platicephaly, closer to that expressed by the smallest specimens of *Homo erectus*, while the length is intermediate between those of small sized and larger sized Asian samples. The maximum width at the temporal lobe and at the Broca area, on the contrary, suggest a derived expansion which is typical of the Neandertal lineage. It must be stressed, however, that among Neandertals the maximum breadth is localised at the temporo-parietal area, while in *Homo erectus* it arises more inferiorly, properly in the temporal lobe (Holloway 1980). In Scp.1 it is parieto-temporal on the right and fully temporal on the left. Anyway, if we compare the endocranial diameters with the corresponding external dimensions, the ratio is strictly comparable with that of the Wurmian Neandertals. It may be hypothesised that many of the resemblances between Scp.1 and specimens referred to *Homo erectus* may be the result of the small size and of a size-related morphological pattern (including the strong degree of platicephaly). At the same time, it is interesting to note that width figures (derived traits) and relative endocranial dimensions of length and height approach Scp.1 to the other Neandertals more than the absolute values.

Whether the results of the geometric morphometric analysis reported here will be confirmed by a larger comparative effort (Bruner, n.d.), it should be concluded that the variability of the endocast architecture among archaic representatives of the genus *Homo* – as far as the mid-sagittal projection of the superior profile is concerned – appears strongly dependent from cerebral dimensions. Size results correlated to 80% of the variance in shape. This allometric pattern seems to be shared by both *Homo erectus* and the Neandertals. We must assume that allometric trend is not only related to overall morphology, but it is also linked to the functional organisation of the brain system, including the expression of circumvolutions, venous sinuses and meningeal patterns.

In light of this analysis of the shape, Scp.1 shows – once more – features that are comparable with the range of variability of *Homo erectus*, sharing with specimens referred to this *taxon* both the extent of brain size and a marked platicephaly. In future dataset modern profiles will also be included giving a more precise results, in particular regarding the role of the parietal areas, which in modern populations are extremely developed. It must be stressed that parietal enlargement has a geometrical effect on the whole cerebral structure. As parietal areas increase, the functional cerebral axis (i.e. the orientation of the brain) changes, pulling down and forward the occipital, temporal and cerebellar lobes. This process involves a rotation of the rear structures under the parieto-temporal areas, while the temporal poles are pushed below the frontal ones. Thus, in *Homo erectus* the cerebral axis is fronto-occipital, whereas among modern humans the axis is fronto-parietal. This must be taken into account when an evolutionary perspective and an analytical context and formalisation of the geometry of the brain are considered (Bruner, n.d.). The role of a posterior parietal cortex development has been previously described as one of the principal factor related to the origin of human evolution itself (Holloway, 1995); it is probably associated to the perception of spatial relationships, processing of visual, auditory and somatic information, as well as to social communication.

Allometry is often the result of ontogenetic and small regulative variations, and it is generally not related to any deep reorganisation of the genome (Gould 1966; Shea 1992; Klingenberg 1998). Given the piece of evidence added so far by the study of the endocast of Saccopastore, and if we consider that specimens as diverse – both in terms of chronology and geography, but also in terms of morphology – as Scp.1 and Trinil 2 may share the same model of brain development, we should assume that evolutionary stasis occurred along the *Homo* lineages, at least before the appearance of modern humans.

Acknowledgements

Sincere thanks go to Giancarlo Gualdi and Stefano Caprasecca (Università di Roma 'La Sapienza', Policlinico 'Umberto I'), for fundamental help and technical assistance in CT-data recording; to Karl Lafaut (Materialise Company in Leuven, Belgium) for his constant attendance to this project; to Juan Luis Arsuaga and Patricio Dominguez (Universidad Complutense de Madrid, Spain), as well as to Fred Spoor (University College London, UK), for useful discussions, suggestions, and advises.

Author address:

Dipartimento di Biologia Animale e dell'Uomo

Università di Roma 'La Sapienza'

Piazzale Aldo Moro 5, 00185 Roma, Italy

* Corresponding author: GM, <giorgio.manzi@uniroma1.it>

BIBLIOGRAPHY

ARSUAGA, J.L., MARTÍNEZ, I., GRACIA, A. & LORENZO, C. 1997, The Sima de los Huesos crania (Sierra de Atapuerca, Spain). A comparative study. *Journal of Human Evolution* 33, p. 219-281.

BOOKSTEIN, F. L., 1991, *Morphometric tools for landmark data.* New York: Cambridge University Press.

BOOKSTEIN, F.L., SCHÄFER, K., PROSSINGER, H., SEIDLER, H., FIEDER, M., STRINGER, C.B., WEBER, G.W., ARSUAGA, J.L., SLICE, D.E., ROHLF, F.J. RECHEIS, W. MIRIAM, A.J. & MARCUS, L.F., 1999, Comparing frontal cranial profiles in archaic and modern Homo by morphometric analysis. *The Anatomical Record (New Anat.)* 257, p. 217-224.

BRADSHAW, J.L., 1988, The evolution of human lateral asymmetries: new evidence and second thoughts. *Journal of Human Evolution* 17, p. 615-637.

BRUNER, E., (n.d.). Computed tomography and paleoneurology: the Saccopastore Neandertals (unpublished manuscript).

CONDEMI, S., 1992. *Les Hommes Fossiles de Saccopastore et leur Relations Phylogénétiques.* Paris: CNRS Editions.

CONROY, G. & VANNIER, M., 1984, Noninvasive three-dimensional computer imaging of matrix-filled fossil skulls by high-resolution computed tomography. *Science* 226, p. 456-226.

FALK, D., 1993, Meningeal arterial patterns in great apes: implications for hominid vascular evolution. *American Journal of Physical Anthropology* 92, p. 81-97.

FELSENSTEIN, J., 1989, PHYLIP: Phylogeny Inference Package (ver. 3.2). *Cladistics* 5, p. 164-166.

GRIMAUD-HERVÉ, D., 1997, *L'évolution de l'encéphale chez Homo erectus et Homo sapiens.* Paris: CNRS Editions.

GOULD, S.J., 1966, Allometry and size in ontogeny and phylogeny. *Biological Review* 41, p. 587-640.

KLINGENBERG, C.P., 1998, Hetrochrony and allometry: the analysis of evolutionary change in ontogeny. *Biological Reviews of the Cambridge Philosophical Society* 73, p. 79-123.

HOLLOWAY, R.L., 1980, Indonesian "Solo" (Ngandong) endocranial reconstructions: some preliminary observations and comparisons with neandertal and Homo erectus group. *American Journal of Physical Anthropology* 553, p. 285-295.

HOLLOWAY, R.L., 1981a, Volumetric and asimmetry determinations on recent hominid endocasts: Spy I and Spy II, Djebel Ihroud I, and the Salé Homo erectus specimen. With some notes on Neandertal brain size. *American Journal of Physical Anthropology* 55, p. 385-393.

HOLLOWAY, R.L., 1981b, The Indonesian Homo erectus brain endocasts revisited. *American Journal of Physical Anthropology* 55, p. 503-521.

HOLLOWAY, R.L., 1995, Toward a synthetic theory of human brain evolution. In *Origins of the Human Brain,* edited by J.P. Changeaux & J. Chavaillon. Oxford: Clarendon Press, p. 42-54.

HOLLOWAY, R.L. & M.C., DE LA COSTE-LAREYMONDIE, 1982, Brain endocast asymmetry in pongids and hominds: some preliminary findings on the paleontology of cerebral dominance. *American Journal of Physical Anthropology* 58, p. 101-110.

HUBLIN, J.J., 1998, Climatic changes, paleogeography, and the evolution of the Neandertals. In *Neandertals and modern humans in Western Asia,* edited by T. Akazawa, K. Aoki & O. Bar-Yosef . New York: Plenum Press, p. 295-310.

LYNCH, J. M., WOOD, C. G. & LUBOGA, S. A., 1996, Geometric morphometrics in primatology: craniofacial variation in Homo sapiens and Pan troglodytes. *Folia Primatologica* 67, p. 15-39.

MANZI, G. & P., PASSARELLO, 1991, Anténéandertaliens et Néandertaliens du Latium (Italie Centrale). *L'Anthropologie* 95, p. 501-522.

MANZI, G., BRUNER, E., CAPRASECCA, S., GUALDI, G. & P., PASSARELLO, 2001, CT-scanning and virtual reproduction of the Saccopastore Neandertal crania. *Rivista di Antropologia* 79, p. 61-72.

MARCUS, L.F., BELLO, E. & GARCÍA-VALDECASAS, A., 1993, *Contributions to Morphometrics.* Madrid : Museo Nacional de Ciencias Naturales.

MARCUS, L.F., CORTI, M., LOY, A., NAYLOR, G.J.P. & SLICE, D., 1996, *Advances in Morphometrics.* New York: Plenum Press (NATO ASI series).

PAGE, R. D. M., 1996. TREEVIEW: an application to display phylogenetic trees on personal computer. *Computer Applied Biosciences* 12, p. 357-358.

PENIN, X., 2000. Applied Procustes Softwares ver. 2.3.

RECHEIS, W., WEBER, G., SCHAFER, K., PROSSINGER, H., KNAPP, R., SEIDLER, H. & ZUR NEDDEN, D., 1999, New methods and techniques in Anthropology. *Collegium Amtropologicum* 23, p. 495-509.

ROHLF, F.J. & BOOKSTEIN, F.L., 1990, *Proceedings of the Michigan Morphometrics Workshop.* Ann Arbor: University of Michigan, Museum of Zoology.

ROHLF, F. J. & MARCUS, L. F., 1993, A revolution in morphometrics. *Tree* 8, p. 129-132.

ROHLF, F. J., 1993, Relative Warp analysis and an example of its application to mosquito wings. In *Contributions to Morphometrics*, edited by L. F. Marcus, E. Bello & A. García-Valdecasas. Madrid : Museo Nacional de Ciencias Naturales, p. 131-159.

ROHLF, F.J., 1997, TpsSplin ver. 1.15. Ecology and Evolution, SUNY at Stony Brook, NY.

ROHLF, F.J., 1998a, TpsRelw ver. 1.18. Ecology and Evolution, SUNY at Stony Brook, NY.

ROHLF, F.J., 1998b, TpsDig ver. 1.2. Ecology and Evolution, SUNY at Stony Brook, NY.

ROHLF, F.J., 2000a, TpsRegr ver. 1.24. Ecology and Evolution, SUNY at Stony Brook, NY.

ROHLF, F.J., 2000b, TpsPLS ver. 1.08. Ecology and Evolution, SUNY at Stony Brook, NY.

SABAN, R., 1995, Image of the human fossil brain: endocranial casts and meningeal vessels in young and adult subjects. In *Origins of the Human Brain*, edited by J.P. Changeaux & J. Chavaillon. Oxford: Clarendon Press, p. 11-38.

SEIDLER, H., FALK D., STRINGER, C., WILFING, H., MULLER, G.B., ZUR NEDDEN, D., WEBER, G.W., REICHEIS, W. & ARSUAGA, J.L., 1997, A comparative study of stereolithographically modelled skulls of Petralona and Broken Hill: implications for future studies of Middle Pleistocene hominid evolution. *Journal of Human Evolution* 33, p. 691-703.

SERGI, S., 1929, La scoperta di un cranio del tipo di Neanderthal presso Roma. *Rivista di Antropologia* 28, p. 457-462.

SERGI, S., 1944, Craniometria e craniografia del primo Paleantropo di Saccopastore. *Ricerche di Morfologia* 20-21, p. 733-791.

SERGI, S., 1948, L'uomo di Saccopastore. *Paleontographia Italica* 42, p. 25-164.

SHEA, B.T., 1992, Developmental perspective on size change and allometry in evolution. *Evolutionary Anthropology* 1, p. 125-134.

SPOOR, F., ZONNEVELD, F. & MACHO, G., 1993, Linear measurements of cortical bone and dental enamel by Computed Tomography: applications and problems. *American Journal of Physical Anthropology* 91, p. 469-484.

SPOOR, F., JEFFERY, N. & ZONNEVELD, F., 2000, Imaging skeletal growth and evolution. In *Development, growth and evolution* edited by P. O'Higgins & M. Cohn. London: Academic Press, p. 124-161.

ZOLLIKOFER, C., PONCE DE LEON, M.S & MARTIN, R.D., 1998, Computer assisted paleoanthropology. *Evolutionary Anthropology* 6, p. 41-54.

ZOLLIKOFER, C. & PONCE DE LEON, M.S., 2000, The brain and its case: computer based case studies on the relation between software and hardware in living and fossil hominid skulls. In *Humanity from African Naissance to Coming Millenia*, edited by P.V. TOBIAS, M.A. RAATH, J. MOGGI-CECCHI, G.A. DOYLE. Firenze – Johannesburg: Firenze University Press & Witwatersrand University Press, p. 379-384.

MODELS OF SHAPE VARIATION BETWEEN AND WITHIN SPECIES AND THE NEANDERTHAL TAXONOMIC POSITION: A 3D GEOMETRIC MORPHOMETRICS APPROACH BASED ON TEMPORAL BONE MORPHOLOGY

Katerina HARVATI

Résumé: La position taxonomique des Néandertaliens est assez controversée. Les espèces fossiles sont souvent définies par comparaison avec des espèces actuelles. Une telle comparaison doit prendre en compte la variation morphologique parmi les populations d'une même espèce, ainsi qu'entre des espèces différentes. Plusieurs caractères Néandertaliens se trouvant sur l'os temporal, deux modèles de variation de la morphologie de l'os temporal ont été développés en utilisant la morphométrie géométrique tridimensionelle. Le premier modèle est basé sur la variation parmi les populations d'hommes modernes, et le deuxième sur la variation parmi les espèces et sous-espèces de chimpanzés. 15 points ont été enregistrés sur l'os temporal de : 12 Néandertaliens, 2 hommes anatomiquement modernes, 4 Européens du Paléolithique supérieur, 2 spécimens du Pléistocène moyen et 270 spécimens de H. sapiens actuel, réprésentant 9 populations, chacune de 30 sujets. L'échantillon des chimpanzés comprenait 35 Pan paniscus, 29 Pan t. troglodytes et 30 Pan t. schweinfurthii. Les spécimens ont été superposés en utilisant le logiciel GRF-ND et les méthodes de superposition Procruste (Generalized Procrustes Analysis). Pour la morphologie de l'os temporal, la distance Mahalanobis entre les Néandertaliens et les populations d'hommes modernes est plus grande que celle entre les deux espèces de chimpanzés. Les Néandertaliens ne montrent aucune similitude morphologique avec les Européens du Paléolithique supérieur et les Européens actuels. Bien que les données des populations humaines modernes se recouvrent largement, il existe de nets groupes géographiques.

Abstract: The taxonomic position of Neanderthals is a matter of wide disagreement. Species recognition in paleontology must be based on analogy with living species, in which both intra- and inter-specific morphological variation is assessed. As several traits that characterize Neanderthals are located on the temporal bone, two models of temporal bone variation were developed using 3D geometric morphometrics, based on modern human populations and chimpanzee species and subspecies. 15 temporal bone landmarks were recorded on 12 Neanderthals, 2 early anatomically modern humans, 4 Late Paleolithic Europeans, 2 Middle Pleistocene specimens and 270 recent humans, the latter representing nine populations of 30 individuals each. The chimpanzee sample included 35 Pan paniscus, 29 Pan t. troglodytes and 30 P. t. schweinfurthii. The specimens were superimposed in GRF-ND using Generalized Procrustes Analysis. Neanderthals are more distant in Mahalanobis distance in their temporal bone morphology from any modern human population than the two chimpanzee species are from each other. They do not show similarities to either modern or Late Paleolithic Europeans. Although the modern groups overlap extensively, they do show geographic clustering.

INTRODUCTION

The taxonomic position of Neanderthals and their role in the evolution of modern humans are at the center of one of the most heated debates in paleoanthropology today. Some authors recognize this fossil group as a different species, *H. neanderthalensis*, and see no evidence of a Neanderthal contribution to the evolution of modern humans in Europe (e.g. Stringer et al. 1984; Stringer 1989, 1994; Tattersall 1986). Other researchers, however, see Neanderthals as a subspecies, or population, of *H. sapiens,* which contributed to some degree to the evolution of modern Europeans (Wolpoff 1989, 1992; Wolpoff et al. 2001). Several intermediate positions have also been formulated, including replacement with various degrees of gene flow from Neanderthals (Bräuer 1992; Duarte et al. 1999). Most authors agree that assignment of species taxa in paleontology must be made based on analogy to living biological species that are phylogenetically, geographically and ecologically similar to the fossil organisms studied (Shea et al. 1993; Szalay 1993). The range of morphological variation within living species must be evaluated, so that a measure of the geographic, sexual and individual variation to be expected in a fossil sample can

be obtained. However, the morphological difference between closely related species must also be assessed when assigning fossil samples to species taxa, as it has been proposed that closely related primate species cannot be differentiated on the basis of bony morphology alone (Tattersall 1986, 1993; Kimbel and Rak 1993).

This study focused on the temporal bone, where many proposed Neanderthal traits are located, such as the small mastoid process, large juxtamastoid eminence, elevated position of the external acoustic meatus, robusticity of the zygomatic process and more coronal orientation of the petrotympanic crest (Vallois 1969; Santa Luca 1978; Hublin 1988; Condemi 1992; Elyaqtine 1996). Most of these traits are difficult to measure directly with traditional caliper measurements, and have not been subject to rigorous quantitative analysis. Furthermore, the temporal bone makes part of the basicranium, which is thought to be very conservative and little affected by epigenetic factors (Olson 1981) and that was recently shown by Wood and Lieberman (2001) to exhibit low levels of intraspecific variation, and are thus well-suited to interspecific comparisons. The objectives of this study were two-fold: a) to obtain measures of variation both within and between species and to apply them to a comparison between Neanderthals and modern humans, and, b) to evaluate the proposed Neanderthal traits quantitatively, using geometric morphometrics (Harvati 2001a, b, in review, in prep). The predictions used were formulated based on two hypotheses. Hypothesis A, that Neanderthals represent a different species from modern humans, predicts that the morphological distance between Neanderthals and modern humans would be greater than the morphological distance between two modern human populations. It would also be greater than that between the two chimpanzee subspecies, and it would be equivalent to that between the two chimpanzee species. Furthermore, Neanderthals would not show morphological similarities to the Late Paleolithic and recent European specimens. Hypothesis B, that Neanderthals represent a subspecies of *H. sapiens*, predicts that the morphological distance between Neanderthals and modern humans would be equivalent to that between any two modern human populations, or that between the two chimpanzee subspecies. It would be smaller than that between the two chimpanzee species. Furthermore, Neanderthals may not show affinities to recent Europeans, but they would show similarities to the Late Paleolithic European specimens (Relethford 2001a).

MATERIALS AND METHODS

Two hundred seventy modern human crania were digitized, representing nine populations of 30 individuals each and spanning the extremes of the modern human geographical range (Table 1), following Howells' seminal study (1973, 1989). Ninety four chimpanzee specimens were also measured, representing the two chimpanzee species, *P. troglodytes* and *P. paniscus*, as well as two subspecies of the common chimpanzee, *P. t. troglodytes* and *P. t. schweinfurthii* (Table 1). The fossil sample comprised twelve Neanderthal specimens from Europe and the Near East; the early Neanderthal specimen from Reilingen; the Middle Pleistocene African specimen Kabwe; two early anatomically modern humans from the Near East; and four Late Paleolithic anatomically modern humans from Europe (Table 2). Where the original fossils were unavailable, casts from the Anthropology Department of the American Museum of Natural History were measured.

The data were collected and analyzed using 3-dimensional geometric morphometrics. The use of geometric

Table 1 - List of specimens by population and sex for the modern human and chimpanzee samples.

Group	Male	Female	Undetermined	Total
Modern Humans	143	126	1	270
Andamanese (And. Islands, India)	13	17	---	30
Australians (New S. Wales, S. Aus.)	19	11	---	30
Berg (Austria)	15	15	---	30
Dogon (Mali, West Africa)	15	15	---	30
Epipaleolithic (Morocco, Algeria)	18	12	---	30
Inugsuk (Greenland)	15	15	---	30
European (Egypt, Dalmatia, Greece, Italy, Germany)	17	12	1	30
San-Hottentot (South Africa)	16	14	---	30
Tolai (New Britain, Melanesia)	15	15	---	30
Chimpanzees	52	40	2	94
Pan paniscus (Zaire)	16	19	---	35
Pan t. schweinfurthii (Zaire)	18	12	---	30
Pan t. troglodytes (Zaire, Cameroon)	18	9	2	29

Table 2 - Fossil human specimens included in the analysis.

Neanderthal	Late Paleolithic	Early anat. modern	Middle Pleistocene
Saccopastore 2	Cro Magnon 1 (cast)	Skhul 5	Kabwe
La Chapelle (cast)	Mladec 2	Qafzeh 9	Reilingen
La Ferrassie 1 (cast)	Predmosti 3 (cast)		
Shanidar 1 (cast)	Predmosti 4 (cast)		
Circeo 1			
Amud 1			
La Quina 27			
Gibraltar 1			
Krapina 39-1			
Krapina C			
Spy 1			
Spy 2			

morphometrics presents several advantages over traditional morphometrics: a) the geometric relationships are preserved, b) visualization of shape changes in specimen space are possible, and most importantly, c) geometric morphometrics enables the quantification of features that cannot be measured with traditional caliper measurements and are therefore usually described qualitatively (Rohlf 1990; Rohlf and Marcus 1993; Dean D. 1993; Slice 1996; O'Higgins and Jones 1998).

The data were collected in the form of 3-D landmark coordinates, using the Microscribe 3DX portable digitizer. Minimal reconstruction was allowed during data collection for specimens where very little damage was observed in the area of interest. Fifteen homologous landmarks were collected on the temporal bone (Table 3) and the landmark coordinates were processed using Procrustes Superimposition (Generalized Procrustes Analysis), using GRF-ND and Morpheus (Slice 1992, 1994-1999).

This method optimally superimposes specimens so that the sum of squares of residuals across specimens and landmarks is minimized, and also removes size differences, so that the differences they exhibit are due to 'shape' (Rohlf 1990). Since reflection of right and left side is possible in GRF-ND, it was possible to combine in one sample fossil specimens preserving the temporal bone on different sides. Missing data were further reconstructed by mirror imaging for right-left homologous landmarks.

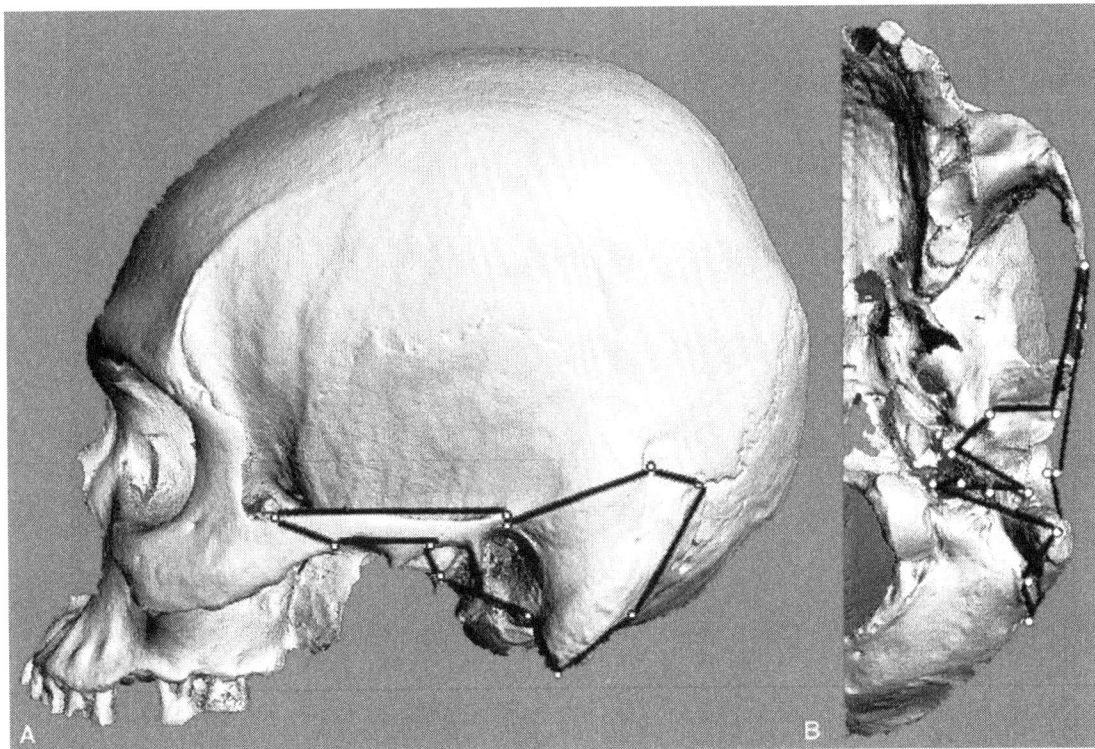

Figure 1 - Temporal bone landmarks, shown on a modern human skull. While dots represent landmarks, black lines between landmarks are links used for convenience in visualization. A: Lateral view, B: Ventral view.

Table 3 - Landmarks measured on the temporal bone.

Temporal bone landmarks	
1. Asterion	(Steps 1-2)
2. Stylomastoid Foramen	(Steps 1-2)
3. Most medial point of the jugular fossa	(Steps 1-2)
4. Most lateral point of the jugular fossa	(Steps 1-2)
5. Lateral origin of the petro-tympanic crest	(Steps 1-2)
6. Most medial point of the petro-tympanic crest at the level of the carotid canal	(Steps 1-2)
7. Porion	(Steps 1-2)
8. Auriculare	(Steps 1-2)
9. Parietal Notch	(Steps 1-2)
10. Mastoidiale	(Steps 1-2)
11. Most inferior point on the juxtamastoid crest (following Hublin 1978a)	(Steps 1-2)
12. Deepest point of the lateral margin of the articular eminence (root of the articular eminence)	(Steps 1-2)
13. Suture between the temporal and zygomatic bones on the inferior aspect of the zygomatic process	(Step 1)
14. Suture between the temporal and zygomatic bones on the superior aspect of the zygomatic process	(Step 1)
15. Most inferior point on the entoglenoid pyramid	(Steps 1-2)

The fitted coordinate configurations resulting from these procedures are thought to lie in Kendall's shape space (Rohlf 1996), although recently Slice (2001) found that they lie in a hemispherical variant of this shape space. As in both cases shape space is non-Euclidean, a projection of these coordinates to tangent space is usually recommended for statistical analysis. However, since biological data are restricted in their variation, the shape space coordinates are almost identical to their projections in tangent space (Slice 2001).

This assumption was tested using TPSSMALL (Rohlf 1998), which compares the Procrustes distances to the Euclidean distances. The correlation between the two distances was very strong (correlation 0.9998, root MS error 0.0004), and the statistical analysis was performed on the fitted coordinates themselves. These were analyzed using principal components analysis (PCA), canonical variates analysis (CVA), Mahalanobis D^2, cluster analysis and minimum spanning tree analysis. An analysis of variance (ANOVA) was performed on the PCA scores to determine the significance of population effects along each component.

The analysis proceeded in two steps using different numbers of landmarks, in order to maximize the number of specimens included. The group membership information used in the CVA was population membership rather than species or genus information, so as not to bias the results toward separation of pre-designated species. Furthermore, the Mahalanobis distance matrices obtained were calculated correcting for unequal sample sizes (Sarmiento and Marcus 2000), and were used to produce cluster and minimum spanning trees. Singletons were excluded from the last two analyses, as their distances from other groups may be overestimated.

RESULTS

Principal Components Analysis

In both steps of analysis modern humans and Neanderthals were separated along PC 1 (41.07% of the total variance in step 1, 43.8 % in step 2, Figure 2a). Chimpanzees were also very widely separated from modern humans along this component. The separation between Neanderthals and modern humans was more evident in the second step, where more Neanderthal specimens were included. However, one Neanderthal, Amud 1, consistently fell within the modern human range along this component. All chimpanzee groups were found to be significantly different from all modern human populations along this component, while Neanderthals were significantly different from all chimpanzee and all modern human populations, including the Late Paleolithic and anatomically modern human specimens. PC 1 was most highly influenced by differences in the position of the tip of the mastoid process, the tip of the juxtamastoid eminence, the lateral end of the petrotympanic crest and porion. The shape differences reflected by this component are the larger size of the mastoid process, the more lateral position the juxtamastoid eminence, the more posterior placement of the lateral end of the petrotympanic crest, and the more medial position of porion in humans relative to chimpanzees. The same shape differences characterize Neanderthals relative to modern humans, but to a lesser extent. The position of Amud 1 well within the modern human range probably reflects the large size of the mastoid process in this specimen. Neanderthals were partially separated from modern humans along a few additional principal components, which, however, reflected a much smaller proportion of the total variance. These included PCs 9 and 15 (Step 1) and PC 10 (Step 2). Taken together, the shape differences that they showed include a more lateral placement of auriculare, the more inferior and posterior position of the root of the articular eminence and the more inferior and medial position of the tip of the juxtamastoid eminence in Neanderthals relative to modern humans. A more extensive analysis of the shape differences between modern human and Neanderthal temporal bones is presented elsewhere (Harvati 2001b, in prep).

Canonical Variates Analysis

As in the PCA, the first canonical variate separated modern humans from chimpanzees and Neanderthals from modern humans (Figure 2b). It explained 68.9 % (Step 1) and 68.6 % of the total variance and in both cases it is most highly

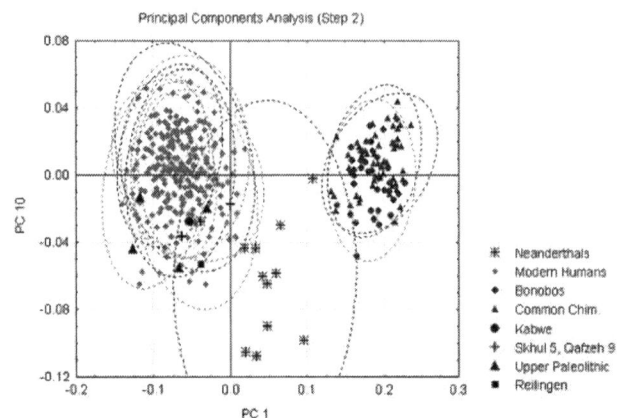

Figure 2 A - Principal Components Analysis (Step 2), PCs 1 and 10.

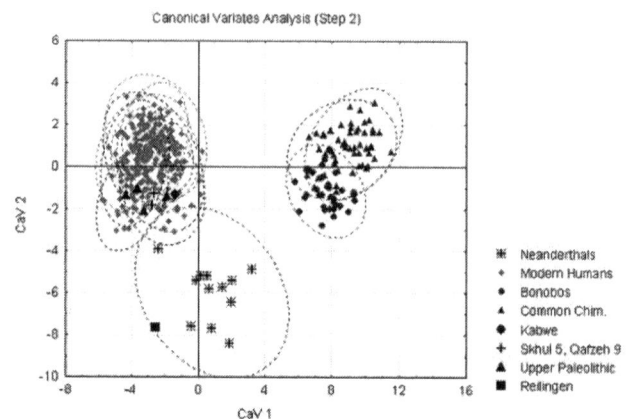

Figure 2 B - Canonical variates analysis (Step 2), CaVs 1 and 2. Dotted lines represent the 95% confidence ellipses for each population.

influenced by PC 1, reflecting similar shape differences to PC 1. Neanderthals were also completely separated from modern humans along CaV 2 (Step 2), which explained 6.5 % of the total variance and was most strongly influenced by PCs 10, 7 and 5.

Mahalanobis D², Cluster Analysis and Minimum Spanning Trees

Neanderthals were found to be more distant in Mahalanobis distances from chimpanzees than they are from modern humans. They were more distant from modern humans than the two chimpanzee species and subspecies were from each other. They were also more distant from all modern human groups than any modern human population was from another. The group that was closest to Neanderthals was the Inugsuk Eskimo population. However, the distance between this population and Neanderthals is very large and probably does not indicate any particular morphological affinities. Neither the recent European groups nor the Late Paleolithic European sample, were found to be close to Neanderthals in Mahalanobis distance. Among recent human groups, the three geographic pairs included (Berg and Europeans, San and Dogon, Australian and Tolai) are closest neighbors to each other in both steps of analysis, showing strong geographic clustering. These results were very similar in both Step 1 and 2, and are summarized for Step 2 by the UPGMA analysis and the minimum spanning tree (Figure 3). They are discussed in greater detail elsewhere (Harvati 2001b, in review).

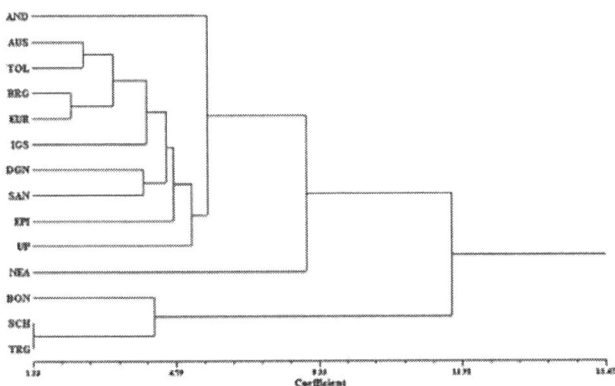

Figure 3 A - Cluster analysis (UPGMA).

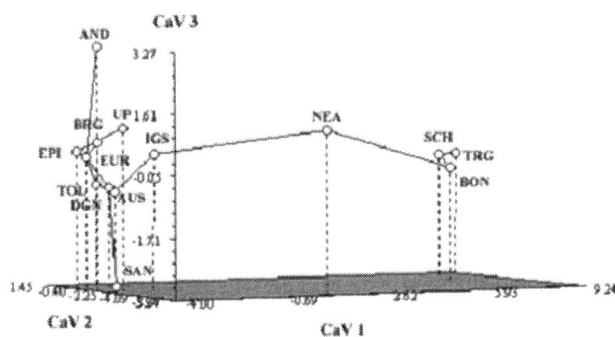

Figure 3 B - Minimum spanning tree (Step 2).

SUMMARY AND CONCLUSIONS

The temporal bone traits measured here were very successful in separating Neanderthals from modern humans, confirming previous qualitative observations. Neanderthals were found to be more distant from modern humans than the two chimpanzee species are from each other, or than any two modern human populations are from each other. Furthermore, they did not show any morphological affinities either to recent Europeans or to the Late Paleolithic European specimens. The morphology of the temporal bone therefore supports the hypothesis that Neanderthals represent a different species from modern humans. Analysis of additional anatomical areas will shed further light on this question.

Acknowledgments

I thank the organizers of this symposium, Bertrand Mafart and Hervé Delingette. I also thank Bill Howells, Chris Stringer, Louise Humphrey, Rob Kruszynski, Yoel Rak, George Koufos, Jakov Radovcic, Henry de Lumley, Marie-Antoinette de Lumley, Dominique Grimaud-Hervé, Gabriella Spedini, Giorgio Manzi, Roberto Macchiarelli, Luca Bondioli, Patrick Semal, the family of Prof. Max Loest, Reinhard Ziegler, Maria Teschler, André Langaney, Mario Chech, Horst Seidler, Sylvia Kirchengast, Niels Lynnerup, Wim van Neer, Ross MacPhee, Ian Tattersall, Gary Sawyer, Ken Mowbray, Eric Delson, Les Marcus, Michelle Singleton, David Reddy and Martin Friess for their help. This research was supported by NSF, the American Museum of Natural History, and the Onassis and CARE Foundations. NYCEP morphometrics contribution No. 6.

Author's address:

Department of Anthropology,
New York University,
25 Waverly Place, New York, NY 10003.
Tel.: (212) 998-8576.
Email: katerina.harvati@nyu.edu

BIBLIOGRAPHY

BRÄUER, G.,1992, Africa's place in the evolution of *Homo sapiens*. In *Continuity or Replacement: Controversies in Homo sapiens Evolution*, edited by G. Bräuer & F.H. Smith, Rotterdam: A. A. Balkema, pp. 83-98.

CONDEMI, S.,1992, Les Hommes Fossiles de Saccopastore et leurs Relations Phylogénétiques. Cahiers de Paléoanthropologie, Paris: CNRS Éditions.

DEAN, D.,1993, The middle Pleistocene *Homo erectus / Homo sapiens* transition: New evidence from space curve statistics. Ph.D. Dissertation, City University of New York, New York.

DUARTE, C., MAURICIO, J., PETTITT, P. B., SOUTO, P., TRINKAUS, E., VAN DER PLICHT, H. & ZILHAO, J. (1999). The early Upper Paleolithic human skeleton from the Abrigo do Lagar Velho (Portugal) and modern human emergence in Iberia. *Proc. natl. Acad. Sci.* **96**: 7604-7609.

ELYAQTINE, M., 1996, L'os temporal chez *Homo erectus* et *Homo sapiens*: Variabilité et évolution. *Revue d'Archaeométrie* **20**: 5-22.

HARVATI, K.,2001a, Models of shape variation within and among species and the Neanderthal taxonomic position: a 3-D geometric morphometric approach on temporal bone morphology. *J. hum. Evol. (Abstracts of the Paleoanthropology Society Meeting)* **40**: A9-A10.

HARVATI, K.,2001b, The Neanderthal problem: 3-D geometric morphometric models of cranial shape variation within and among species. PhD thesis, City University of New York.

HARVATI, K. in review. The Neanderthal taxonomic position: models of intra- and inter-specific morphological variation.

HARVATI, K. in prep. Quantitative analysis of Neanderthal and modern human temporal bone morphology.

HOWELLS, W. W.,1973, *Cranial Variation in Man: A Study by Multivariate Analysis of Patterns of Difference Among Recent Human Populations*. Papers of the Peabody Museum of Archaeology and Ethnology, Harvard University, Vol. **67**.

HOWELLS, W. W.,1989, *Skull Shapes and the Map: Craniometric Analyses in the Dispersion of Modern Homo*. Papers of the Peabody Museum of Archaeology and Ethnology, Harvard University, Vol. **79**.

HUBLIN, J. J., 1988, Caractères dérivés de la région occipitomastoïdienne chez les Néandertaliens. *L'Anatomie*, L'Homme de Néandertal, vol. **3**: 67-73.

KIMBEL, W. H & Rak, Y. (1993). The importance of species taxa in paleoanthropology and an argument for the phylogenetic concept of the species category. In *Species, Species Concepts and Primate Evolution*, edited by W. H. Kimbel & L. B. Martin, New York: Plenum Press, pp. 461-484.

O'HIGGINS, P., JONES, N., 1998, Facial growth in *Cercocebus torquatus*: an application of thrre-dimensional geometric morphometric techniques to the study of morphological variation. *J. Anat.* **193**: 251-272.

OLSON, T. R., 1981, Basicrania and evolution of the Pliocene hominids. *In Aspects of Human Evolution*, edited by C. B. Stringer, London: Taylor and Francsis, , pp. 99-128.

RELETHFORD, J. H., 2001a, Absence of regional affinities of Neanderthal DNA with living humans does not reject multiregional evolution. *Am. J. phys. Anthrop.*, **115**: 95-98.

ROHLF, J. F., 1998 (Copyright). TpsSmall, version 1.18.

ROHLF, J. F., 1990. Rotational fit (Procrustes) methods. In *Proceedings of the Michigan Morphometrics Workshop,* edited by F. J Rohlf & F. L. Bookstein, Ann Arbor: The University of Michigan Museum of Zoology, pp. 227-236.

ROHLF, J. F. & MARCUS, L. F., 1993, A revolution in morphometrics. *Trends in Ecology and Evolution* **8**: 129-132.

SANTA LUCA, A. P., 1978, A re-examination of presumed Neanderthal-like fossils. *J. hum. Evol.* **7**:619-636.

SARMIENTO, E. E. & MARCUS, L. F., 2000, The os navicular of humans, great apes, OH 8, Hadar and Oreopithecus: Function, phylogeny and multivariate analyses. *Am. Mus. Novit.* **3288**: 1-38.

SHEA, B. T., LEIGH, S. T. & GROVES, C. P. ,1993, Multivariate craniometric variation in chimpanzees: Implications for species identification. In *Species, Species Concepts and Primate Evolution*, edited by W. H. Kimbel & L. B. Martin, New York: Plenum Press , pp. 265 -296.

SLICE, D.E.,1992, 1994 (Copyright). GRF-ND: Generalized rotational fitting of n-dimensional landmark data.

SLICE, D. E., 1994-1999, (Copyright). Morheus et al.

SLICE, D. E.,1996, Three-dimensional generalized resistance fitting and the comparison of least-squares and resistant fit residuals. In *Advances in Morphometrics,* edited by L. F. Marcus, M. Corti, A. Loy, G. J. P. Naylor, D. E. Slice, NATO ASI Series. New York: Plenum Press, pp.179-199.

SLICE, D. E., 2001, Landmark coordinates aligned by Procrustes analysis do not lie in Kendall's shape space. *Systematic Biology* 50: 141-149.

STRINGER, C. B., 1989, The origin of early modern humans: a comparison of the European and non-European evidence. In *The Human Revolution*, edited by P. Mellars & C. Stringer Princeton, Princeton University Press, pp. 232-244.

STRINGER, C. B.,1994, Out of Africa - A personal history. In *Origins of Anatomically Modern Humans*, edited by M.H. Nitecki & D. V. Nitecki, Plenum Press, New York, pp. 149-172.

STRINGER, C. B., HUBLIN, J. J. & VANDERMEERSCH, B., 1984, The origin of anatomically modern humans in Western Europe. In *The Origins of Modern Humans: A World Survey of the Fossil Evidence*, edited by F. H. Smith & F. Spencer, New York: Liss, pp. 51-135.

SZALAY, F. S., 1993, Species concepts: The tested, the untestable and the redundant. In *Species, Species Concepts and Primate Evolution*, edited by W. H. Kimbel & L. B. Martin, New York: Plenum Press, pp. 523-535.

TATTERSALL, I., 1986, Species recognition in human paleontology. *J. hum. Evol.* **15**: 165-175.

TATTERSALL, I. ,1993, Speciation and morphological differentiation in the genus *Lemur* In *Species, Species Concepts and Primate Evolution*, edited by W. H. Kimbel & L. B. Martin, New York: Plenum Press, pp. 163-176.

VALLOIS, H. V.,1969, Le temporal Néanderthalien H 27 de La Quina. Étude anthropologique. *L'Anthropologie* **73**: 524-400.

WOLPOFF, M. H.,1989, Multiregional evolution: The fossil alternative to Eden. In *The Human Revolution,* edited by P. Mellars & C. Stringer Princeton: Princeton University Press, pp. 62-108.

WOLPOFF, M. H.,1992, Theories of modern human origins. In *Continuity or Replacement: Controversies in Homo sapiens Evolution*, edited by G. Bräuer & F. H. Smith, Rotterdam: A. A. Balkema, pp. 25-63.

WOLPOFF, M. H., HAWKS, J., FRAYER, D. & HUNLEY, K., 2001, Modern human ancestry at the peripheries: A test of the replacement theory. *Science* **291**: 293-297.

WOOD, B. & LIEBERMAN,D.,2001, Craniodental variation in *Paranthropus boisei*: A developmental and functional perspective. *Am. J. phys. Anthrop.*, **116**: 13-25.

THE USE OF 3D LASER SCANNING TECHNIQUES FOR THE MORPHOMETRIC ANALYSIS OF HUMAN FACIAL SHAPE VARIATION

Martin FRIESS, Leslie F. MARCUS†, David P. REDDY & Eric DELSON

Résumé: Cette étude a pour objectif de montrer d'intérêt de l'utilisation du scanner à laser pour l'analyse de la morphologie crâniofaciale chez l'homme actuel et fossile. Nous présentons un protocole de mesure permettant l'acquisition de données surfaciques que nous avons appliqué à l'étude de possibles adaptations crâniofaciales au froid dont nous proposerons également une modélisation. Jusqu'à présent, nos résultats sont compatibles avec le concept d'une voûte crânienne fonctionnant comme un radiateur dont le rapport volume-surface varie conformément aux prédictions dérivées de la règle de Bergmann., tant pour les hommes actuels que fossiles. La surface relative de la face varie à l'opposé des prédictions dérivées de la règle d'Allen. Ceci nous amène à suggérer que la morphologie faciale observée chez les Inuits ou les Néandertaliens résulte de facteurs qui ne sont pas en rapport avec des conditions climatiques.

Abstract: The present study explores the application of laser surface scanning to the analysis of craniofacial morphology in living and fossil humans. We present a measurement procedure for the assessment of relative surface areas and apply it to examine possible craniofacial cold adaptations, for which we also present a theoretical model. In it's current state of progress, our analysis supports the idea that the braincase functions as a radiator, and that its volume-to-surface area ratio varies consistently with predictions derived from Bergmann's rule, for living human populations as well as for Neandertals. The relative surface area of the face is found to vary opposite to predictions derived from Allen's rule. This suggests that the facial morphology seen among Inuit or Neandertal populations is driven by factors that are mainly unrelated to climatic conditions.

INTRODUCTION

The morphology of the human facial skeleton has long been of interest to anthropologists and other researchers. Its particular architecture seen in Neandertals has, for decades, initiated numerous attempts to seek functional interpretations in terms of either masticatory stress or cold adaptation. Several authors have also tried to identify similar cause-and-effect relationships among human populations living under harsh climatic conditions, such as Eskimos and Fueguians.

The continuing rise of 3D imaging and measuring techniques and their application in anthropology provide new possibilities of studying functional morphology (e.g. Lyons et al. 2000). The purpose of this paper is to explore the use of 3D laser surface scanning techniques for the quantitative analysis of the human craniofacial skeleton. A special emphasis is put on morphometric aspects of cold adaptation. In the following, we propose a measurement protocol for the assessment of volume/area ratios in order to test whether the skull exhibits features that are consistent with hypothesized thermoregulatory constraints. We will also present preliminary results of this approach.

REVISITING COLD ADAPTATION

The assumption that the human face may show signs of cold adaptation in certain populations, like Neandertals and or Inuit, goes back to Hrdlièka (1910). He suggested that a narrow nasal aperture, as seen in Inuit, may increase temperature and humidity of the inspired air, an idea that was later supported by Wolpoff (1968) and Hylander (1977). The apparent contradiction with the proposed cold adaptation of

the relatively broad Neandertal nose has been discussed by Hylander (1977).

Coon (1950) proposed that facial flatness and protruding malars, another feature considered to be common among Inuit, is functionally adapted to cold temperatures. This hypothesis refers implicitly to the notion that relative surface area ought to be rather small and therefore less exposed under cold climatic conditions. However, the idea was rejected by Steegman (1970, 1972) on experimental grounds. He showed that the temperature measured at the surface of the malar bone is not significantly different from that at any other facial region when exposed to cold. In the 1970 study he observed that "mongoloid" malars are larger, and that better warming occurs in the malar in Europeans rather than in Asians. In the 1972 study, he observed a temperature decrease to be correlated with malar protrusion: A protruding malar (a "mongoloid" feature) becomes colder, as it is more exposed. Regarding the thermoregulation of the braincase, Steegman (1970) suggested that the relative surface area should be smaller in cold adapted populations in order to reduce heat loss.

Finally, in a somewhat different approach to thermoregulation, Dean's model (1988) assumes that large nasal cavities and faces are advantageous for exposure to hot climates. Following this model one would expect that, inversely, small nasal cavities and small faces have an adaptive advantage in regions of cold climate.

RATIONALE

Summarizing the existing models on thermoregulation leads to the conclusion that all of them cannot be true, as they are in part contradictory: We feel that it is most reasonable to assume that any thermoregulatory process, if observable in the exoskeleton, would have to follow the rules of climatic adaptation according to Bergmann and Allen (Rensch, 1936). As discussed by Ruff (1991), the phenomenon that generally underlies cold adaptation is that an organism changes it's surface area to volume ratio in order to reduce the chances of heat loss. Therefore, the appropriate theoretical model for craniofacial cold adaptations is the following:

1. The surface area of the braincase relative to brain volume is reduced in populations exposed to cold temperatures (Bergmann's rule).

2. The surface area of the face and/or the zygomatic bone has the properties of an extremity. Its surface area relative to the skull is reduced (Allen's rule)

MATERIAL

Data were sampled from populations representing 11 different geographic regions, covering a wide range of climatic conditions (table 1). All specimens are part of the collections of the AMNH. For each population, 3 adult individuals of both sexes were sampled, yielding a total of 66 individuals. Only individuals with an erupted M3 or a fused sphenobasilar synchondrosis were considered. The sex of each individual was taken from records for most of the cases and, if not available, estimated by one of us (MF) on the basis of general cranial features. Information about the specimens' geographic origin was available from Museum records, although for some individuals this information was rather vague. For instance, all Australian skulls are reportedly from "South Australia", a province that extends roughly over 10 degrees of latitude.

In addition, mean annual temperatures were recorded for the geographic origin of each sample according to meteorological observations available online. The recorded temperatures vary from -25°C to +30°C.

We also included a series of fossil hominins, mainly Neandertals, whose facial skeletons were sufficiently well preserved or assessable through reconstruction. These comprise: La Ferrassie, La Chapelle aux Saints, Guattari 1,

Shanidar 1, Amud 1, Kabwe, and Steinheim. All fossil data were obtained from high quality casts in the Department of Anthropology of the AMNH.

LASER SCANNING TECHNIQUE

The data were acquired with a Cyberware 3030 RGB, a table-sized 3D laser scanner with optional color registration. The device yields a complete 3D virtual model of the visible surface of a given object. The accuracy of the obtained data ranges between 0.25 and 0.5 mm (Frieß et al. in press).

A Laser scanner projects a light beam whose reflection from a moving object is captured by a CCD. With the device used for this study, the object is placed on a turntable that performs a linear movement during which the laser beam projects a single line onto the surface, thereby producing a scanning profile of the object from one specific viewpoint (fig. 1). Rotating the turntable and repeating the scan yields a series of profiles from a number of viewpoints determined by the user, which therefore allow a computational reconstruction of the object via triangulation. Subsequent manual editing steps are required to ascertain clean raw data sets.

Once the specimen is completely scanned and the raw data have been manually cleaned, the software (headus.com) can be used to obtain linear measurements as well as area and volume of a surface.

Following the theoretical considerations (elaborated above) we measured the surface area of the braincase, the facial skeleton, and the zygomatic bone for each cranium. In addition, we measured the cranial capacity of each individual using millet seeds.

The area measurements were defined as follows (fig. 2):

Braincase: The area of the neurocranium most likely to be exposed to external temperature while not being covered by the neck. This includes the complete external frontal (exluding the roof of the orbit) and parietal bones, the squamous portion of the occipital and the temporal (above the zygomatic arch)

Table 1 - Mean annual temperatures recorded for each geographic origin of the sample

Geographic origin	N	Mean annual temperature	Geographic origin	N	Mean annual temperature
Greenland "Eskimos" Smith Sound	6	-9.6	Mexico, Tarraso	6	16.1
Alaska "Eskimos" Point Barrow	6	-12.6	Thailand, Bankok	6	27.9
Mongolia	6	10	Egypt, Gizah	6	21.3
Patagonia	6	11.2	Tanzania, Pare & Ungueno	6	26.3
Tierra del Fuego	6	5.6	South Africa, Khoisan	6	15.9
			South Australia	6	25.4

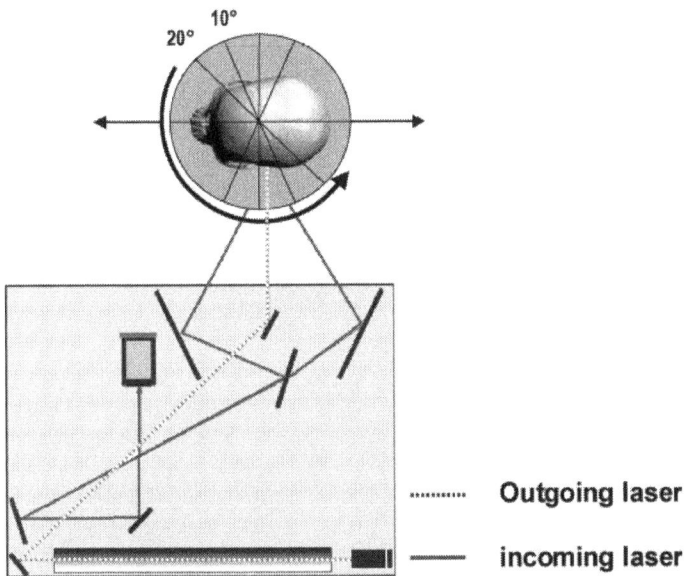

Figure 1 - Scheme of the Cyberware 3030 laser scanner (seen from above), illustrating the way it captures the surface of an object.

Facial and Neurocranial surface area measurements

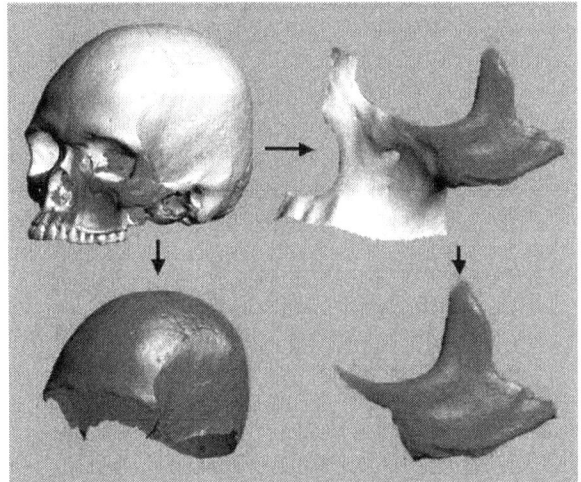

Figure 2 - Surface area measurements taken on the braincase and the face

as well as the external surface of the great wing of the sphenoid surrounded by the temporal, the frontal and the zygomatic.

Face: The external, anterior and lateral surface area of the zygomatic and maxillary bone above the alveolar arch. Measures of the maxilla were limited to the portion above the alveolar arch and anterior to the zygomatic root, considering that these are the portions most likely to be exposed to external temperature.

Zygomatic: The anterior external surface area of the zygomatic bone (thus a subset of the previous area).

Despite their potential, or at least suggested role for thermoregulation, we decided to exclude the nasal bones from the measurements. They are, frequently broken or completely absent among modern humans, and very rarely preserved in fossil hominins.

PRELIMINARY RESULTS

The overall degree of variation of the neurocranial surface area relative to cranial capacity is rather small among the different modern human samples (fig. 3). Significant differences are observed at both ends of the range, with Greenland Eskimos exhibiting the smallest relative surface area and Australian Aborigines the largest. The relative surface area (scaled to cranial capacity) of the Australian sample is certainly related to, but not fully caused by, the relatively small cranial capacity of the sample. The raw surface area of this Australian sample is also absolutely larger than the area in any other sample.

Grouping the samples together in 2 large sets, roughly equivalent to warm versus cold climate populations (see table 1 for annual temperatures) reveals a significant decrease of relative surface area with decreasing annual temperature.

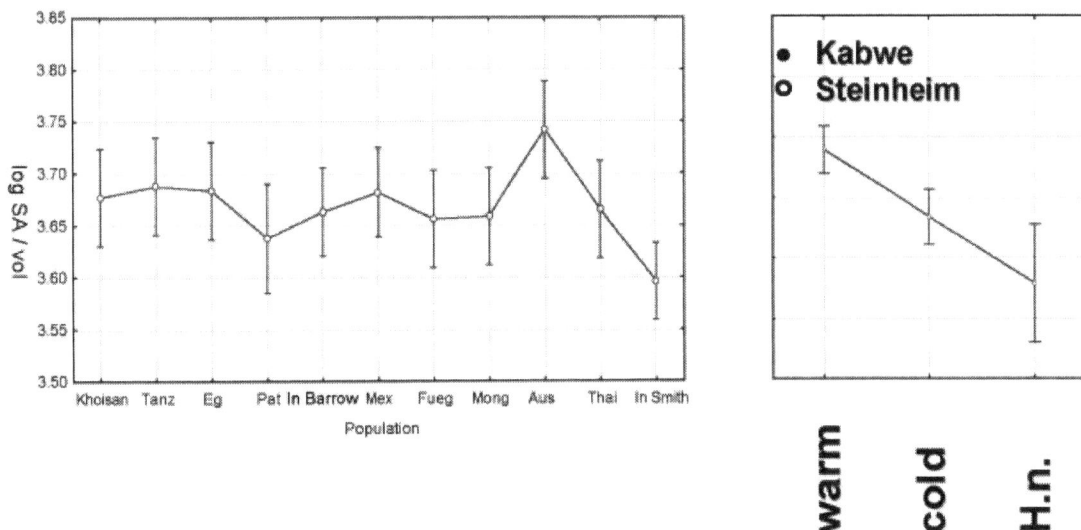

Figure 3 - Variation among extant and extinct humans of the neurocranial surface area relative to cranial capacity. Left: variation across geographic samples. Right: variation as a function of mean annual temperature. The position of fossils reflects only the relative surface area (ordinate).

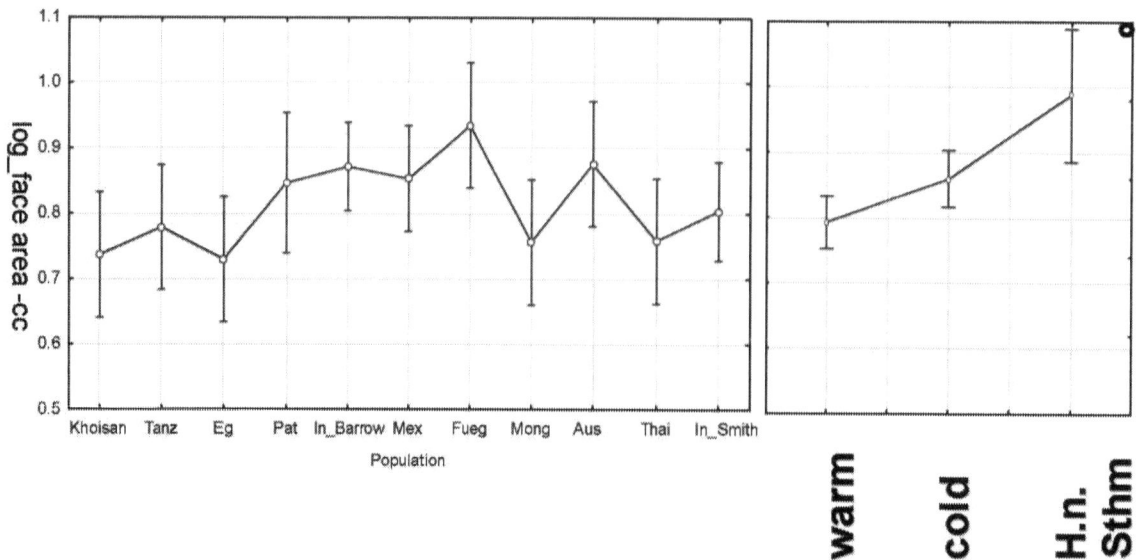

Figure 4 - Variation among extant and extinct humans of the facial surface area relative to the overall cranial surface area. Left: variation across geographic samples. Right: variation as a function of mean annual temperature. The position of fossils reflects only the relative surface area (ordinate).

Furthermore, when data from fossil hominins are compared to the scores of living humans, the result indicates that Neandertals have an even smaller relative surface area of their braincase. On the other hand, the Kabwe and Steinheim fossils exceed the modern human range of relative neurocranial surface area. Kabwe comes from an equatorial African environment, while Steinheim is generally considered to derive from a fully interglacial central European habitat comparable to those of the present day.

The facial surface area relative to the overall surface (face plus braincase) area shows the same basic pattern of variation (fig 4). While all groups taken separately show a rather heterogeneous picture, pooling warm climate populations on the on hand and cold climate populations on the other reveals a clear trend toward an increase in relative facial surface area with decreasing temperature. Data for all fossil specimens exceed the relative facial surface area of moderns, regardless of which hominid group they represent.

DISCUSSION

Given the methodological scope of this paper, one of it's major contributions is to provide a model approach to the study of volume-to-surface ratios and their functional significance in terms of thermoregulation. The technical advantages of a 3D laser scanning system are quite evident for the assessment of surface area data, which otherwise are accessible only through approximate equations using linear dimensions (Steegmann 1972). The obvious need for this type of data can be conveniently filled with this new technology.

In terms of functional considerations, the present study can provide preliminary insight into the volume-to-surface area ratios involved. The data presented so far suggest that the human head, and the braincase in particular, is adapted so as to reduce heat loss under cold climatic conditions. This

observation can be made on a global scale, when generally warm climates are compared to generally cold climates. The inclusion of fossil specimens indicates that the relatively large cranial capacity of Neandertals and the associated relative small surface area of the neurocranium are adaptive for heat retention, at least when compared to modern human variation. When compared to older fossils that are not supposed to have lived in a cold climate (Kabwe and Steinheim) the Neandertal braincase still exhibits a relatively small surface area. Additional comparison and further confirmation of such a trend of reduced relative surface area among so caled "classic" Neandertals would ultimately suggest that cold adaptation is not only expressed in Neandertal postcranial morphology, but that the braincase was shaped following similar demands.

The data for the facial portions reveal a somewhat different picture. Contrary to theoretical predictions, modern humans living in a subarctic environment are characterized by relatively large facial surface areas, an observation that is consistent with earlier reports on Inuit facial dimensions (Hrdlička, 1910). Therefore, it has to be assumed, until further analyses are completed, that the relatively large faces of the Inuit and Neandertals are unrelated to climatic conditions.

SUMMARY AND CONCLUSION

The present study investigates new approaches to the study of functional morphology using a laser surface scanner. The device can be conveniently used to access surface area measurements otherwise largely inaccessible with conventional devices. An example for a possible application to paleoanthropology is given with the analysis of volume to surface area ratios of the human skull. A preliminary comparison based on 11 modern human populations as well as Late Pleistocene Neandertals and Middle Pleistocene warmer-climate fossils indicates that the braincase of populations living under subarctic conditions is adapted to reduce heat dissipation. When the face is examined under the same assumption, the data obtained so far are not consistent with models of climatic adaptations *sensu* Allen's rule. Therefore, we should either seek alternative models of cold adaptations or abandon the idea that facial morphology is the result of such adaptations.

Acknowledgements

We are very grateful to Drs. Ian Tattersall and Kenneth Mowbray, and to Gary Sawyer (Department of Anthropology, AMNH) for permitting the use of the collection of modern human skulls and fossil casts, as well as for providing their time and expertise during sampling. We are also indebted to K. Baab, and T. Capellini for their assistance with scanning and data editing as well as for their stoic patience when dealing with chronically underpowered workstations. This research was supported by NSF ACR 9982351 to the American Museum of Natural History. This paper, NYCEP morphometrics contribution number [4], is dedicated to Leslie F. Marcus who died while this manuscript was in print.

Authors' addresses:

Martin FRIESS[1, 3, 4], Leslie F. MARCUS[3, 4, 6], David P. REDDY[4, 5] & Eric DELSON[1, 2, 3, 4]

[1] Department of Anthropology, Lehman College/CUNY, Bronx, NY 10468

[2] Ph.D. Program in Anthropology, CUNY Graduate Center, New York, NY 10016

[3] Division of Paleontology, American Museum of Natural History (AMNH), New York, NY 10024

[4] NYCEP (New York Consortium in Evolutionary Primatology)

[5] Interdepartmental Laboratories, AMNH

[6] Department of Biology, Queens College/CUNY, Flushing, NY 11367

BIBLIOGRAPHY

COON, C.S., GARN, S.M., & BIRDSELL, J.B., 1950, *Races: a study of the problems of race formation in man.* Springfield: Thomas.

DEAN, M.C. 1988, Another look at the nose and the functional significance of the face and nasal mucous membrane for cooling the brain in fossil hominids. *J. Hum. Evol.* 17, p. 715-718.

FRIEß, M., DELSON E., MARCUS L.F. & REDDY D.P., 2002 in press, The use of laser scanning techniques in paleoanthropology: A preliminary evaluation.

HERNANDEZ, M., FOX, C.L. & GARCIA-MORO, C., 1997, Fueguian Cranial Morphology: The Adaptation to a Cold, Harsh Environment. *Am. J. Phys. Anthropol.* 103, p. 103-117.

HRDLIČKA, A., 1910, Contribution to the anthropology of Central and Smith Sound Eskimos. *Anthropological Papers of the American Museum of Natural History.* 5, part 2.

HYLANDER W.L., 1977, The Adaptive Significance of Eskimo Craniofacial Morphology. In *Orofacial growth and Development*, edited by AA Dahlberg & TM Graber. The Hague: Mouton, p. 129-169.

LYONS, P.D.; RIOUX, M. & PATTERSON R.T., 2000, Application of a Three-Dimensional Color Laser Scanner to Paleontology. *Paleontologia Electronica* 3(2), p. 1-16.

RENSCH, B., 1936, Studien über klimatische Parallelität der Merkmalsausprägung bei Vögeln und Säugern. *Arch. Naturg.* 5, 317-363.

RUFF, C.B., 1991, Climate and body shape in hominid evolution. *J. Hum. Evol.* 21, p. 81-105.

STEEGMANN, A.T. JR., 1970, Cold adaptation and the human face. *Am. J. Phys. Anthropol* 32, p. 243-250.

STEEGMANN, A.T. Jr., 1972, Cold Response, Body Form, and Craniofacial Shape in Two Racial Groups of Hawaii. *Am. J. Phys. Anthropol.* 37, p. 193-222.

WOLPOFF, M., 1968, Climatic influence of skeletal nasal aperture. *Am. J. Phys. Anthropol.* 29: 405-42.

3D IMAGE PROCESSING FOR THE STUDY OF THE EVOLUTION OF THE SHAPE OF THE HUMAN SKULL: PRESENTATION OF THE TOOLS AND PRELIMINARY RESULTS

Gérard SUBSOL, Bertrand MAFART, Alain SILVESTRE, Marie-Antoinette DE LUMLEY

Résumé: Nous présentons une méthode automatique qui permet de visualiser et d'analyser, en trois dimensions, l'évolution de la forme du crâne humain à partir d'images scanographiques. Un premier algorithme extrait automatiquement des « lignes de crête » à partir des scanographies d'un crâne d'un Homme moderne et du moulage du crâne d'un Homme préhistorique. Ces lignes correspondent aux lignes saillantes de la surface crânienne. Elles servent de repères à un algorithme de mise en correspondance pour trouver automatiquement les points homologues entre les deux crânes. À partir de ces points appariés, on calcule une transformation de l'espace qui superpose les deux crânes. Il est alors possible de visualiser et d'analyser l'évolution du crâne entre l'Homme préhistorique et moderne. Nous appliquons cette méthode au crâne de l'Homme de Tautavel, daté d'environ 450 000 ans, et nous présentons des applications futures en reconstruction faciale et en analyse morphométrique tridimensionnelle.

Abstract: We present an automatic method that allows one to visualize and analyze, in three dimensions, the evolution of the shape of the human skull from CT-Scan images. A first algorithm automatically extract the "crest lines" from CT-Scan images of the skull of a modern Man and of a cast of a skull of a prehistoric Man. Those lines correspond to the salient lines of the skull surface. They will be used as landmarks to automatically find the homology points between the two skulls. Based on these couple of matched points, we compute a volumetric transformation that superposes the two skulls. It makes possible to visualize and analyze the evolution of the skull between the prehistoric and the modern Man. We have applied this method to the skull of Tautavel Man, dated of about 450,000 years, and we present future applications in facial reconstruction and three-dimensional morphometry.

INTRODUCTION

Computer Tomography Scan images are more and more used in paleoanthropology (Spoor et al., 2000), especially for the study of the craniofacial massif. The fossil, or its mold, is placed into a Computer Tomography device (see Figure 1) and we obtain, in few minutes, a series of several tens of digital images that represent the successive slices of the anatomical structure. These images are, in general, of a resolution of 512 by 512 pixels which are coded in several thousands of gray levels. They are then "stacked" in order to build up a three-dimensional image. CT-Scan devices that are routinely used in medical radiology have a resolution of one millimeter whereas special industrial micro-scanners can reach up to a resolution of one hundred microns (Thompson and Ilerhaus, 1998).

Some image processing algorithms developed for medical imaging (Ayache, 1998) or Computer Assisted Design are applied to extract the surface of the structure from the 3D image and to display it, from any point of view, on the screen of a computer. More generally, these algorithms allows one to interact with the virtual representations of the fossils to study them carefully (Weber, 2001), (Zollikofer et al., 1998). The paleontologist can combine the virtual bone fragments to test different reconstruction hypotheses (Kalvin et al., 95), (Thompson & Illerhaus, 1998, (Braun et al., 1999, Ponce de León & Zollikofer, 1999). New virtual fragments can be created by symmetrizing or adjusting the size of the real fragments. It is also possible to model some taphonomic deformations (Ponce de León & Zollikofer, 1999). The paleontologist is also able to easily visualize the internal structures of the virtual fossils, as the endocranium (Conroy et al., 1990), (Conroy et al., 1998), (Ponce de León & Zollikofer, 1999), the bony labyrinth of the inner ear (Spoor & Zonneveld, 1995, Thompson & Illerhaus, 1998), (Ponce de León & Zollikofer, 1999), the frontal (Thompson & Illerhaus, 1998), maxillary (Rae & Koppe, 2000) or paranasal (Ponce de León & Zollikofer, 1999) sinuses. Morphometry tools allow to take easily measurements that are very complex to obtain in reality, as the thickness of bones (Zollikofer et al., 1998) or the curvature radius of the semi-circular canals (Spoor & Zonneveld, 1995), (Thompson & Illerhaus, 1998). More generally, it becomes possible to perform a complete three-dimensional study of the shape of the structure (Subsol et al., 2000, Ponce de León & Zollikofer, 2001). Finally, real replications of the virtual reconstructions can be obtained by some **virtual prototyping** processes as laser stereolithography (Seidler et al., 1997, Zollikofer et al., 1998).

In this paper, we present a new method based on state-of-the-art image processing algorithms that allow to analyze **automatically**, in three dimensions, the evolution of the shape of the human skull. We follow the scheme proposed by d'Arcy Thompson at the beginning of the last century (Thompson, 1917): first, we compute a 3D deformation function between the fossil we want to study and a reference skull; and then, we use tools to visualize this function and to emphasize the differences between the two skulls.

The main goal of this paper is to assess if the presented automatic image processing tools can be successfully applied. For this purpose, we show some preliminary results that are based on a CT-Scan image of a dry skull of a Modern Man

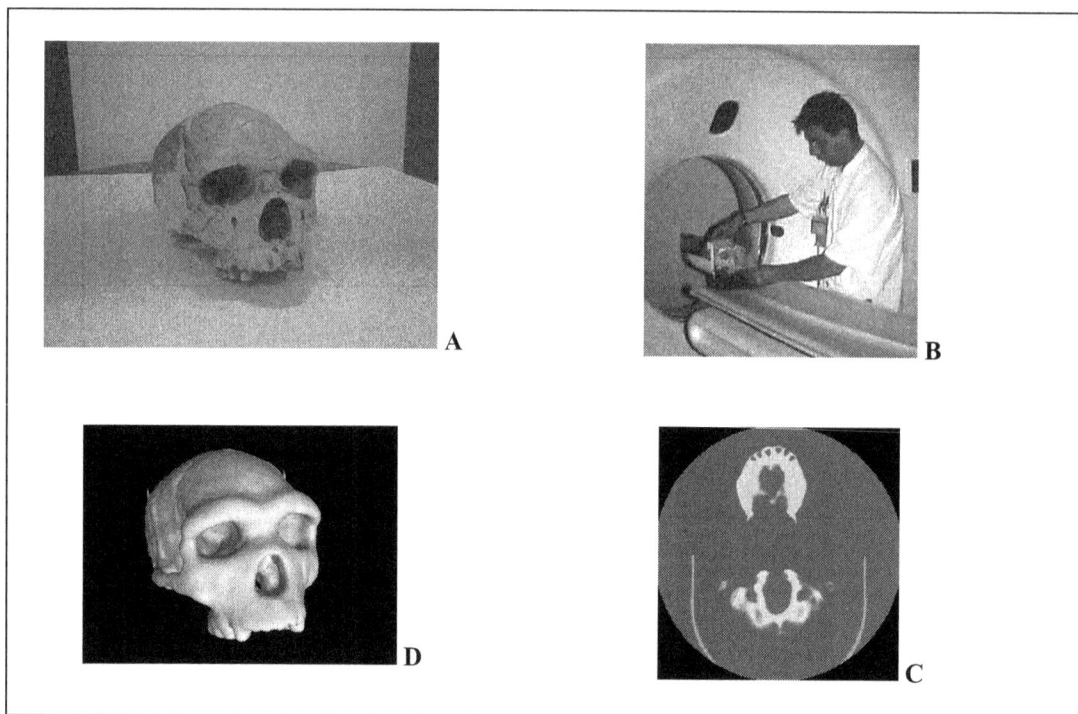

Figure 1 - Obtaining a virtual fossil
The anatomical structure or its mold (A) is placed in the CT-Scan device (B). We obtain then a series of several tens of digital images of 512 by 512 pixels in gray levels that correspond to slices (C). It is then possible to "stack" the slices and to extract the surface of the anatomical structure to visualize it in 3D on a computer screen (D).

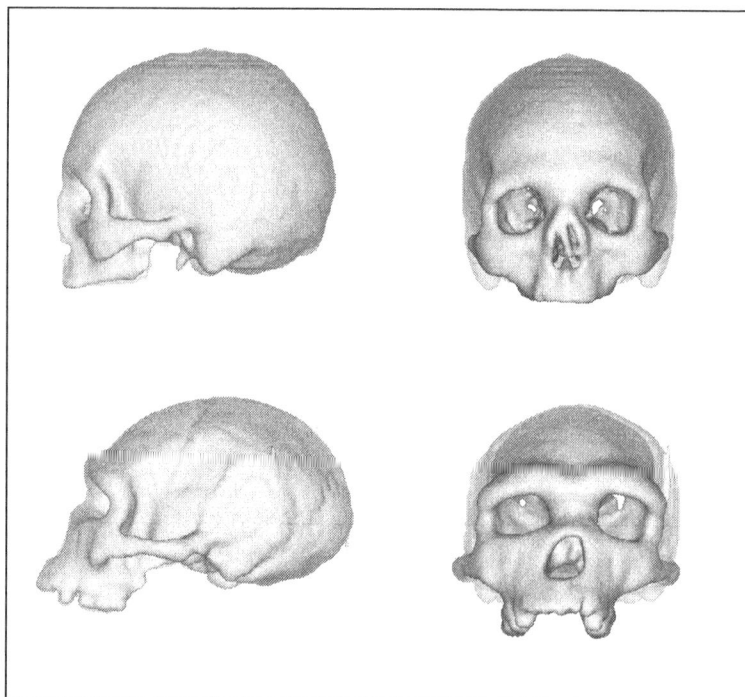

Figure 2 - How to compare the two skulls: Modern Man (up) and Tautavel Man(bottom).
A methodology exists, that consists in computing and analyzing a 3D deformation
function between the two skulls.

(61 slices with a thickness of 3 mm composed of 512 by 512 pixels of 0.6 by 0.6 mm, data by courtesy of Gérald Quatrehomme, University of Nice, France) and a CT-Scan of a mold of the reconstruction of the skull of Tautavel Man (154 slices with a thickness of 1 mm composed of 512 by 512 pixels de 0.5 by 0.5 mm) (see Figure 2). The reconstitution is based on the face (Arago XXI) and the right parietal (Arago XLVII) that were found in the Arago cave at Tautavel in 1971, the left parietal being obtained by symmetry, on a mold of the Swanscombe occipital, and on the temporal bone and its symmetric of Sangiran 17 (Pithecanthropus VIII) (de Lumley, 1982).

PRESENTATION OF THE METHOD

Extraction of Feature Points and Lines

To compute the 3D transformation, we have to find some landmarks on the surface of the skull. They must be defined by an unambiguous mathematical formula to be automatically computed and be anatomically relevant to characterize the structure. We choose **crest lines** (Thirion & Gourdon, 1996, Subsol et al., 1998) that are defined by the extrema of the principal curvature that has the largest absolute magnitude, along its associated principal direction (see Figure 3). Due to their definition, these lines follow the salient lines of a surface. We can verify this in Figure 4 where the crest lines that were automatically extracted in a CT-Scan of the skull of a Modern Man emphasize the mandible, the orbits, the cheekbones or the temples and also, inside the cranium, the sphenoid and temporal bones as well as the foramen magnum.

Salient structures are also used by doctors as anatomical landmarks. For example, the crest lines definition is very close to the **ridge lines** described in (Bookstein & Cutting, 1988) (see Figure 5, left) and that are type II landmark in Bookstein's typology (Bookstein, 1991). In Figure 5, right, we display on the same skull the crest lines (in gray) which were automatically extracted and the ridge lines (in black) which were extracted semi-manually under the supervision of an anatomist. The two sets of lines are very close (Thirion et al., 1996), showing that crest lines would have a strong anatomical significance. Notice that ridge lines have also been used in paleontology to compare the Homo Erectus and the Homo Sapiens (Dean, 1993).

Registration of Feature Lines

We extract 536 crest lines composed of 5,756 points on the skull of the Modern Man and 337 crest lines with 5,417 points on Tautavel Man's skull. Now, we have to find the correspondences between these features (see Figure 6). Usually, this is done manually by an anatomist who knows the **biological homology**: two features are put into correspondence if they characterize the same biological functionality. In our case, there is so many points that this is no more possible and we have to design an algorithm to find correspondences automatically. This is a very well known problem in 3D image processing called **registration** (Ayache, 1998). We have developed a method described in (Subsol, 1995, Subsol et al., 1998) that deforms iteratively and continuously the first set of lines towards the second one in order to superimpose them. At the end of the process, each point P_i of the first set is matched with the point Q_i of the second set that is the closest, and some inconsistent correspondences are discarded. In our example, the algorithm finds in some minutes, on a standard personal computer, 1,532 points pairings (P_i, Q_i). As they are located all around the inside and outside surfaces of the skull, it becomes really possible to analyze the total structure in three dimensions.

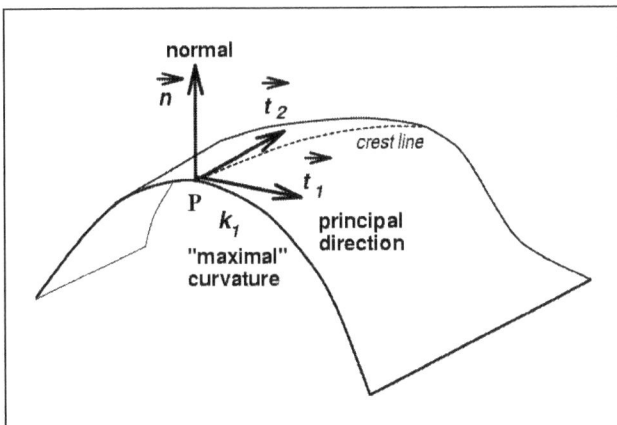

Figure 3 - Mathematical definition of crest lines
- k_1: maximal principal curvature in absolute value
- t_1: associated principal direction
  grad k1.t1=0 $\Leftrightarrow$ P is a crest point

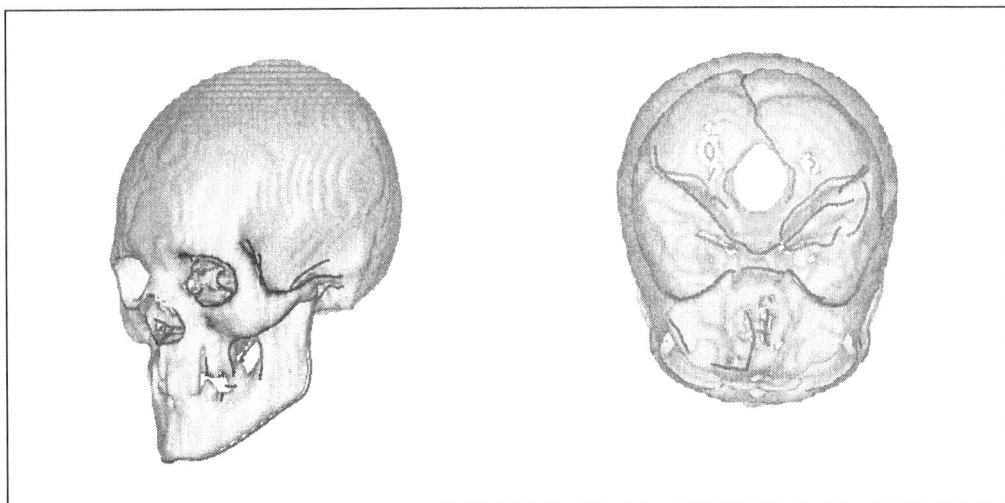

Figure 4 - crest lines automatically extracted in a CT-Scan of the skull of a Modern Man.
Notice how crest lines emphasize the mandible, the orbits, the cheekbones or the temples and also, inside the cranium, the sphenoid and temporal bones as well as the foramen magnum.

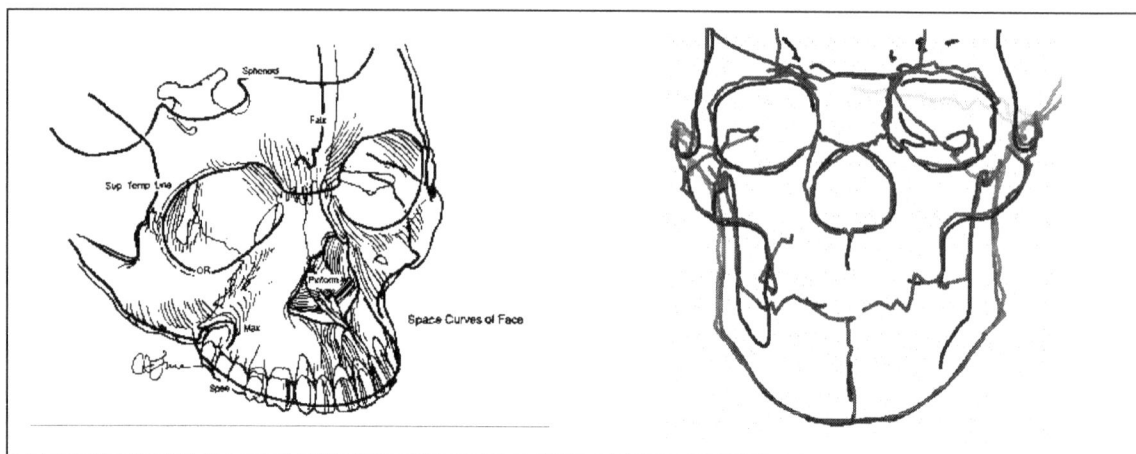

Figure 5 - comparison of ridge lines and crest lines
Left (excerpt from (Bookstein & Cutting, 1988)): ridge lines are extracted semi-manually under the supervision of an anatomist and are used for applications in craniofacial surgery and paleontology.
Right: the superimposition of crest (in gray) and ridge lines (in black) shows that crest lines have a strong anatomical significance, even if they are based on a mathematical definition.

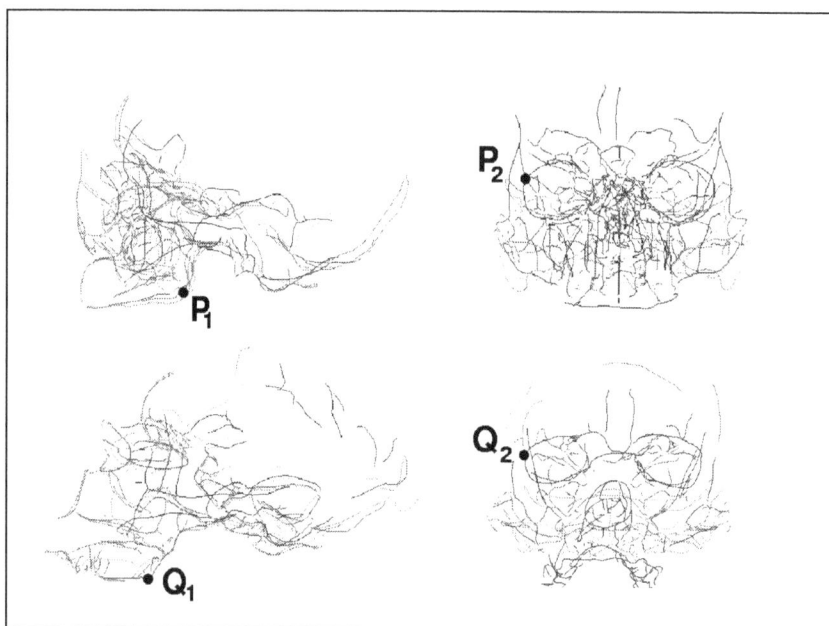

Figure 6 - crest lines automatically extracted and the registration problem. The difficulty is to find the correspondences between these features as, for example, the pairings (P_1, Q_1) or (P_2, Q_2).
Up: crest lines on the skull of the Modern Man (536 lines and 5756 points)
Bottom: crest lines on the skull of Tautavel Man (337 lines and 5417 points).

We checked on several skull data that these registration results are consistent with those obtained by an other automatic method and by a semi-manual method where an anatomist supervises the detection of homologous points (Thirion et al., 1996).

Geometrical Normalization

In ontogenetic and evolutive shape transformation studies, we should not take into account differences of position, orientation and size, since they cannot be considered as true morphological differences. This requires to compute the three following transformations between two specimens to compare: the rotation $\underline{R}$, the translation $\underline{T}$ and the scaling $\underline{s}$.

Several methods exist to compute $(\underline{s},\underline{R},\underline{t})$ based on pairs of homologous points (P_i, Q_i), as the Procrustes superimposition (Boostein, 1991) or the least-square minimization that leads to:

$$(\underline{s},\underline{R},\underline{t}) = \text{Argmin}_{(s,R,t)} \Sigma_i \| sR\ P_i + t - Q_i \|^2)$$

By applying the inverse transformations $(\underline{s}^{-1}, \underline{R}^{-1}, \underline{t}^{-1})$, we can "normalize" the shape of the second skull that becomes comparable to the first one.

Nevertheless, more complex taphonomic transformations modified the shape of the fossils (Ponce de León & Zollikofer, 1999). Thus, in Figure 7 up, we can notice how the Tautavel

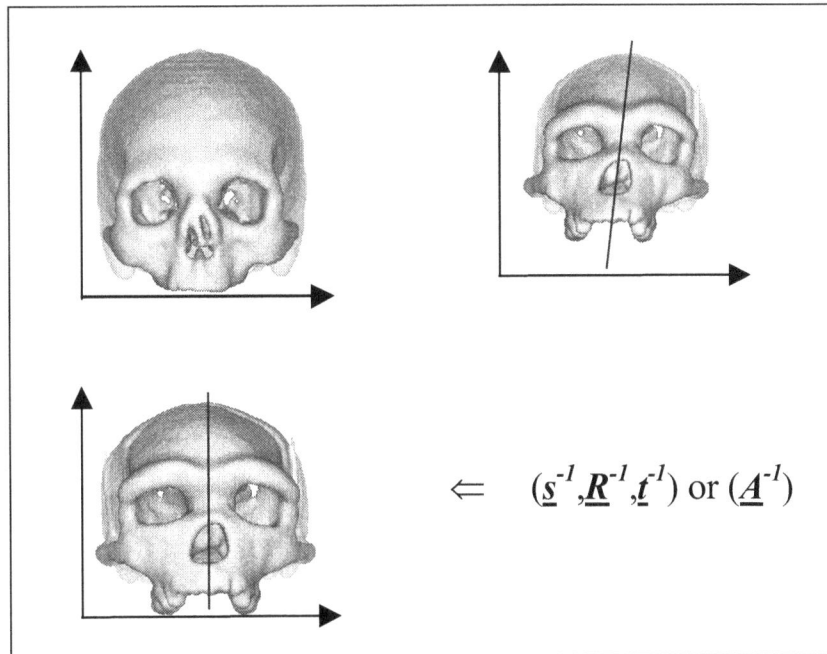

Figure 7 - geometrical normalization of the Tautavel Man's skull
This requires to model and compute not only the differences of position, of orientation and size but also some taphonomic deformations. After the normalization (bottom), the skull becomes comparable to the Modern Man's one (up, left).

Man's skull is bent. This is due to the fact that it was laid on the side and was compressed by the gravity. We have modeled this deformation by an affine transformation $\underline{A}$. We compute it by the least-square method and applied it to he original skull. We can see in Figure 7, bottom, how the skull was rectified and be made comparable to the skull of the Modern Man.

Another way to recover the bending of the skull consists to extract automatically the mid-sagittal plane (Thirion et al., 2000) and to realign it with the vertical plane. Nevertheless, the knowledge of the in situ orientation of the fossil is indispensable, since similar deformations might be the result of different diagenetic events (Ponce de León & Zollikofer, 1999).

Computing the 3D Transformation

Now, we have to compute the 3D transformation between the fossils. The **Thin-Plate Spline** method (Bookstein, 1991), widely used in morphometry, allows to compute such a function that interpolates the displacements of the homologous points (P_i, Q_i) with some mathematical properties of regularity. Nevertheless, interpolation is relevant when the matched points are totally reliable and distributed regularly (for example, with a few points being located manually). In our case, these points are not totally reliable due to possible mismatches of the registration algorithm and are sparse in a few compact areas as they belong to lines. So, we have developed a spline approximation function that is regular enough to minimize the influence of an erroneous matched point (Declerck et al., 1995). The coordinate functions are then computed by a 3D tensor product of B-spline basis

functions. To compute this 3D transformation $\underline{T}$, we maximize the weighted sum of an approximation criterion (quadratic distance between $T(P_i)$ and Q_i) and a regularization criterion (minimization of the second order derivatives that corresponds to the "curvature" of the function):

$$\underline{T} = \text{Argmin}_{(T)} \Sigma_i \|T(P'_i)Q'_i\|^2 + \rho \iiint (\partial^2 T/\partial x^2) + (\partial^2 T/\partial x \partial y) + ..$$

The parameter ρ tunes the approximation accuracy or the smoothness of the transformation.

APPLICATION FOR THE STUDY OF TAUTAVEL MAN'S SKULL

Analysis of the Deformation

We compute the 3D transformation between the Modern Man and Tautavel Man based on the features lines. By applying it to a 3D regular mesh, it is possible to visualize the differences between the two structures. We can notice in Figure 8 that the deformed mesh emphasizes the main features of the Tautavel Man: low skull, receding forehead, protuberant face as well as a an important frontal dissymmetry due to the taphonomic deformations (Mafart et al., 1999).

It is also possible to have a quantitative overview of the deformation. For example, in Figure 9, we colored the surface skull according to the magnitude of the deformation: violet (dark gray) for small displacements and green (light gray) for the largest ones. This shows the importance of the elongation of the face. (Ponce de León & Zollikofer, 2001)

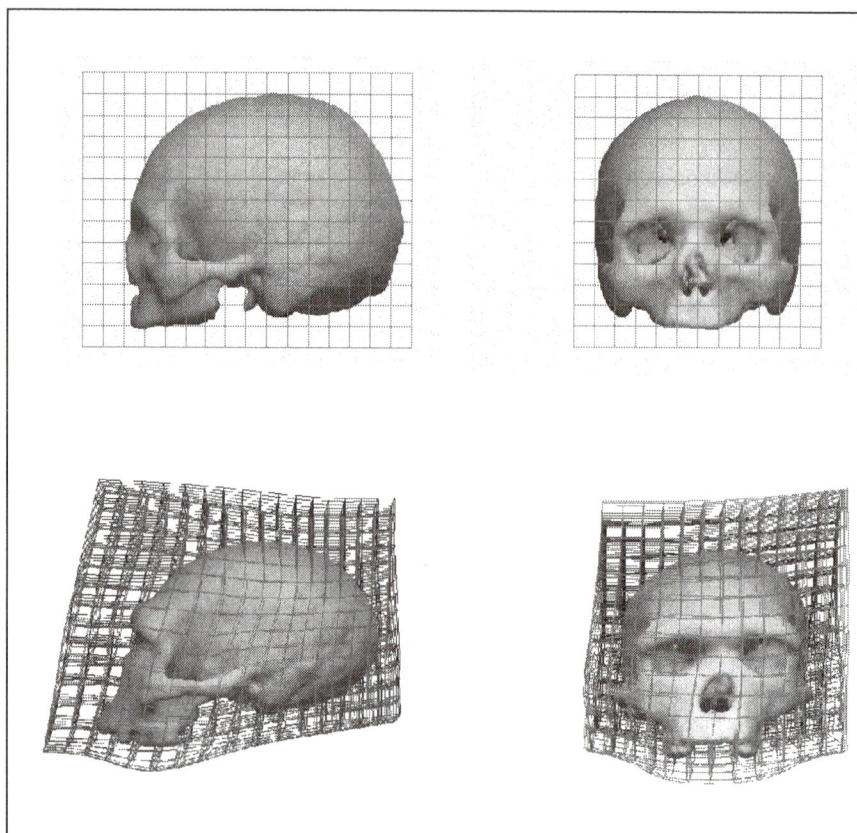

Figure 8 - 3D transformation between the skull of the Modern Man (up) and of Tautavel Man(bottom). Notice how the deformed mesh emphasizes the main features of Tautavel Man: low skull, receding forehead, protuberant face as well as an important frontal dissymmetry due to the taphonomic deformations.

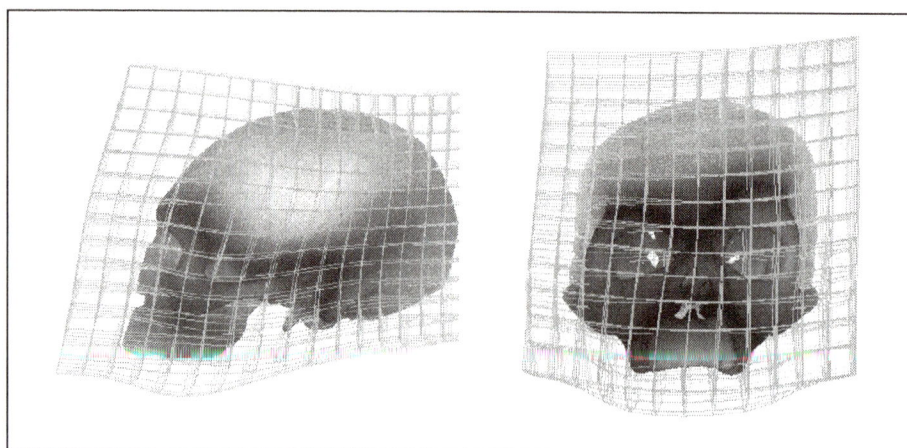

Figure 9 - a quantitative visualization of the 3D deformation.
The color table on the surface skull characterizes the magnitude of the deformation: violet (dark gray) for small displacements and green (light gray) for the largest ones. This emphasizes, in particular, the importance of the elongation of the face

proposes other visualization methods as using colors to indicate the direction of the deformation (inward/outward) or displaying the displacement vector field.

Facial Reconstruction

In (Quatrehomme et al., 1997), we propose an automatic method to perform a 3D facial reconstruction based on the 3D images of an unknown skull and of a reference skull and

face (see Figure 10). We register automatically the 3D images by using crest lines and we compute a 3D transformation between the two skulls. If we assume that the shape of the face follows more or less the shape of the skull, we can apply this 3D deformation to the reference face and infer the face corresponding to the unknown skull.

As a reference face, we use the CT-Scan of the mold of the face of the Modern Man (62 slices with a thickness of 3mm

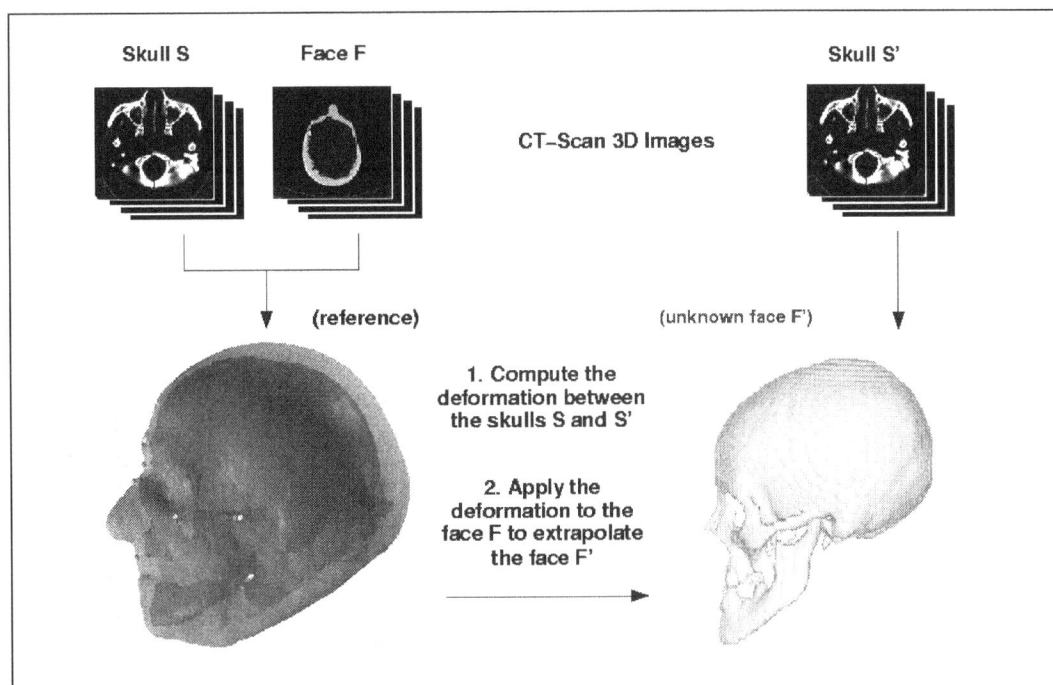

Figure 10 - an automatic scheme for face reconstruction.
The two skulls are registered by using crest lines and the 3D deformation is applied to the reference face to infer the unknown face.

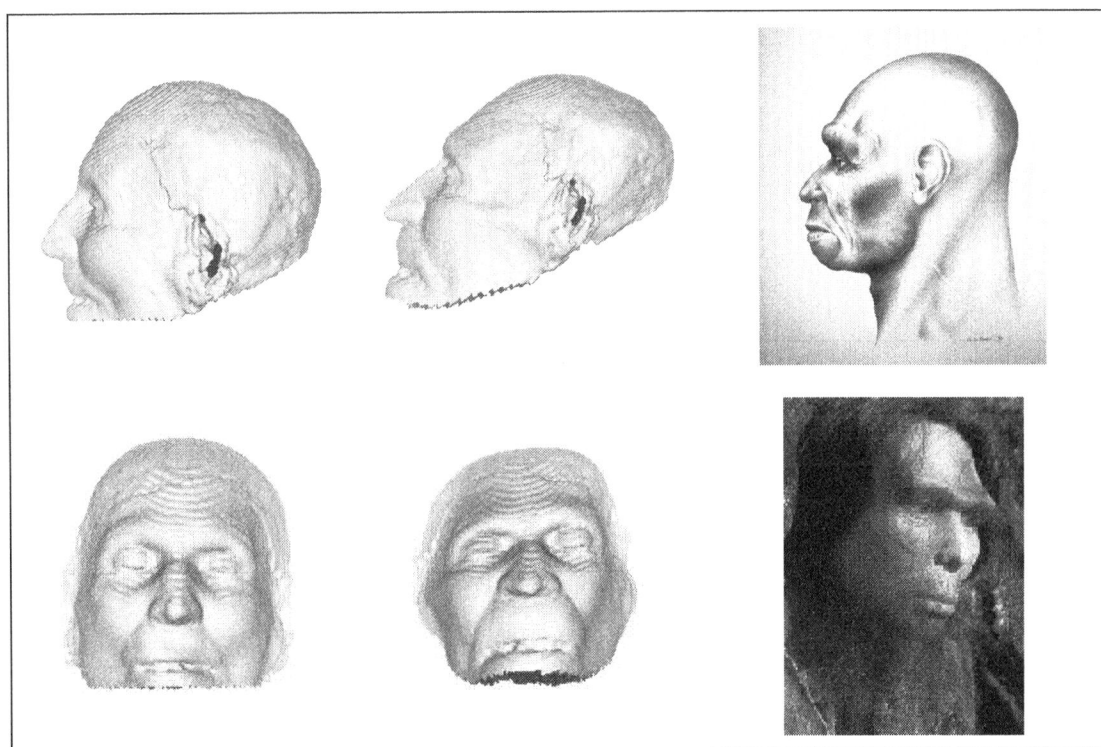

Figure 11 - automatic facial reconstruction of the Tautavel Man's face;
The automatic process deforms the reference face (left) to infer the face corresponding to the skull of Tautavel Man (middle).We can compare this reconstruction to the ones presented on the Web site (right):
http://www.culture.fr/culture/arcnat/tautavel/francais/hominter.htm

composed of 512 by 512 pixels of 0.6 by 0.6mm, data by courtesy of Gérald Quatrehomme, University of Nice, France) that we have aligned manually with the corresponding image of the skull (see Figure 11, left). We can see the result of the automatic reconstruction process in Figure 11, middle. In spite of using a face of an old person, whereas Tautavel Man was quite young, the result appears consistent with other facial reconstructions of his face (see Figure 11, right).

CONCLUSION

In this paper, we have presented several 3D image processing tools – feature extraction, feature registration, complex deformation computation – that can be combined in order to compute and analyze the deformation between two specimens. We have applied an entirely automatic methodology to the study of the shape of the skull of Tautavel Man and we present some preliminary results. Even if they have not yet been compared to the current established paleontology knowledge, we think that they are encouraging and assess the utility of such automatic tools, that are faster than manual procedures, that give reproducible results, and that can be easily parameterized to allow the paleontologist to test several hypotheses.

The development of such tools requires a close collaboration between physicians, anatomists, computer scientists morphometricians and paleontologists. We plan to improve all the steps of the scheme, especially the morphometric analysis. In particular, we will study how to decompose the 3D deformation into a small number of basic and "characteristic" deformations, as "relative warps" (Bookstein et al., 1999) or "principals warps" (Bookstein, 1991, Ponce de León & Zollikofer, 2001). We plan also to apply all this methodology to other anatomical structures as, for example, the human pelvis (Marchal, 2000) or animal bones (Rogers, 1999) and to study pathologic deformations (Mafart, 1998). Endly, we will test our tools with 3D images acquired by new modalities as laser scanning (Kullmer et al., 2001) or Magnetic Resonance Imaging (Steiger, 2001).

Authors' addresses:

Gérard Subsol[1], Bertrand Mafart[2], Alain Silvestre[3], Marie-Antoinette De Lumley[2]

[1] Laboratoire d'Informatique, Université d'Avignon, 339, chemin des Meinajariès, Agroparc B.P. 1228 – 84911 Avignon Cedex 9, France - Gerard.Subsol@lia.univ-avignon.fr

& Projet EPIDAURE, Institut National de Recherche en Informatique et en Automatique, Sophia Antipolis, France

[2] Laboratoire d'Anthropologie, Faculté de Médecine, UMR 6569, Université de la Méditerranée, Marseille, France - bmafart@aol.com,

[3] Hôpital Militaire Robert Picqué, Bordeaux, France

Correspondance to: Gérard Subsol

BIBLIOGRAPHY

AYACHE, N., 1998, L'analyse automatique des images médicales, état de l'art et perspectives. *Annales de l'Institut Pasteur*, vol. 9, no. 1, p. 13-21. Electronic version: http://www.inria.fr/rrrt/rr-3364.html

BOOKSTEIN, F.L., 1991, *Morphometric tools for landmark data*, Cambridge University Press.

BOOKSTEIN, F.L. & CUTTING, C., 1988, A proposal for the apprehension of curving cranofacial form in three dimensions. In K. Vig et A. Burdi, eds., *Cranofacial Morphogenesis and Dysmorphogenesis*, p. 127-140, University of Michigan Press.

BOOKSTEIN, F.L., SCHÄFER, K., PROSSINGER, H., SEIDLER, H., FIEDER, M., STINGER, C., WEBER, G.W., ARSUAGA, J.L., SLICE, D.E., ROHLF, F.J., RECHEIS, W., MARIAM, A.J. & MARCUS, L.F., 1999, Comparing Frontal Cranial Profiles in Archaic and Modern *Homo* by Morphometric Analysis. *The Anatomical Record (New Anat.)*, 257, p. 217-224

BRAUN, M., BOUCHET, P., HUBLIN, J.J. & MALLET, J.L., 1999, Les reconstitutions virtuelles des hommes préhistoriques. *Dossier Pour la Science : Les origines de l'humanité*, p. 78-81.

CONROY, G.C, VANNIER, M.W. & TOBIAS, P.V., 1990, Endocranial Features of Australopithecus africanus Revealed by 2- and 3-D Computer Tomography. *Science*, vol. 247, p. 838-841.

CONROY, G.C., WEBER, G.W., SEIDLER, H., TOBIAS, P.V., KANE, A. & BRUNSDEN, B., 1998, Endocranial Capacity in an Early Hominid Cranium from Sterkfontein, South Africa. *Science*, vol. 280, p. 1730-1731.

DEAN, D., 1993, *The Middle Pleistocene Homo erectus/Homo sapiens Transition: New Evidence from Space Curve Statistics*. Ph.D. Thesis, The City University of New York.

DECLERCK, J., SUBSOL, G., THIRION, J.P. & AYACHE, N., 1995, Automatic retrieval of anatomical structures in 3D medical images. In N. AYACHE, ed., *Computer Vision, Virtual Reality and Robotics in Medicine*, Lecture Notes in Computer Science 905, p. 153-162, Nice (France), Springer. Electronic version: http://www.inria.fr/rrrt/rr-2485.html

KALVIN, A.D., DEAN, D. & HUBLIN, J.J., 1995, Reconstruction of Human Fossils. *IEEE Computer Graphics and Applications*, vol. 15, no 1, p. 12-14.

KULLMER, O., SCHRENK, F., BROMAGE, T., HUCK, M., 2001, (HOTPAD) Hominid Tooth Pattern Data-collection based on optical 3D Topometry. In *Colloquium on 3D Imaging for Paleoantrhopology and Prehistoric Archeology, XIVth Congress of the International Union for Prehistoric and Protohistoric Sciences*. Liège (Belgium).

DE LUMLEY, H., DE LUMLEY, M.A. & DAVID, R., 1982, Découverte et reconstruction de l'Homme de Tautavel. In *Ier congrès de paléontologie humaine*, tome 1, Nice (France).

MAFART, B., 1998, Le crâne romain d'Arles : une syphilis frontale et naso-palatine post-Colombienne, apport des nouvelles méthodes de datation C14 pour la paléopathologie. In *Bulletin et Mémoires de la Société d'Anthropologie de Paris n.s.*, t 10, 3-4, p. 333-344, 1998.

MAFART, B., MÉLINE, D., SILVESTRE, A. & SUBSOL, G., 1999, *3D Imagery and Paleontology: Shape differences between the skull of Modern Man and that of Tautavel Man*. B. Hidoine, A. Paouri, designers and directors, video 451-452, Department of Scientific Multimedia Communication, INRIA. This movie was showed at the exhibition *Homo Erectus à la conquête du monde*, Musée de l'Homme, Paris in 1999-2000.

Electronic version: http://www.inria.fr/multimedia/Videotheque/0-Fiches-Videos/451-fra.html

MARCHAL, F., 2000, A new morphometric analysis of the hominid pelvic bone. *Journal of Human Evolution*, 38, p. 347-365.

PONCE DE LEÓN, M.S. & ZOLLIKOFER, C.P.E., 1999, New Evidence from Le Moustier 1: Computer Assisted Reconstruction and Morphometry of the Skull. *The Anatomical Record*, 254, p. 474-489.

PONCE DE LEÓN, M.S. & ZOLLIKOFER, C.P.E., 2001, Neanderhal cranial ontogeny and its implications for late hominid diversity. *Nature*, vol. 412, p. 534-537.

QUATREHOMME, G., COTIN, S., SUBSOL, G., DELINGETTE, H., GARIDEL, Y., GREVIN, G., FIDRICH, M., BAILET, P. & OLLIER, A., 1997, A Fully Three-Dimensional Method for Facial Reconstruction Based on Deformable Models. *Journal of Forensic Sciences*, 42(4) p. 649-652.

RAE, T.C. & KOPPE, T., 2000, Isometric scaling of maxillary sinus volume in hominoids. *Journal of Human Evolution*, 38 p. 411-423.

ROGERS, S.W., 1999, Allosaurus, Crocodiles, and Birds: Evolutionary Clues from Spiral Computed Tomography of an Endocast. *The Anatomical Record (New Anat.)*, 257, p. 162-173.

SEIDLER, H., FALK, D., STRINGER, C., WILFING, H., MÜLLER, G.B., ZUR NEDDEN, D., WEBER, G.W., REICHEIS, W., ARSUAGA, J.L., 1997, A comparative study of stereolithographically modelled skulls of Petralona and Broken Hill : implications for future studies of middle Pleistocene hominid evolution. *Journal of Human Evolution*, 33, p. 691-703.

SPOOR, F. & ZONNEVELD, F., 1995, Morphometry of the primate bony labyrinth : a new method based on high resolution computed tomography. *Journal of Anatomy*, 186, p. 271-286.

SPOOR, F., JEFFERY, N. & ZONNEVELD, F., 2000, Using diagnostic radiology in human evolutionary studies. *Journal of Anatomy*, 197, p. 61-76. Electronic version: http://evolution.anat.ucl.ac.uk/people/spoor/spmain.htm

STEIGER, T., 2001, Nuclear magnetic resonance imaging in paleontology. *Computers & Geosciences*, 27, p. 493-495.

SUBSOL, G., 1995, *Construction automatique d'atlas anatomiques morphométriques à partir d'images médicales tridimensionnelles.* Ph.D. Thesis, École Centrale Paris. Electronic version: http://www.inria.fr/RRRT/TU-0379.html

SUBSOL, G., THIRION, J.P. & AYACHE, N., 1998, A General Scheme for Automatically Building 3D Morphometric Anatomical Atlases : application to a Skull Atlas. *Medical Image Analysis*, 2(1), p. 37-60.

SUBSOL, G., MAFART, B., MELINE, D., SILVESTRE, A. & DE LUMLEY, M.A., 2000, Traitement d'images scanographiques appliqué à l'étude tridimensionnelle de l'évolution de la forme du crâne humain. In P. Andrieux, D. HADJOUIS, A. DAMBRICOURT-MALASSE, eds., *L'identité humaine en question*, p. 92-101. Collection Paléoanthropologie et paléopathologie osseuse, Artcom.

THIRION, J.P. & GOURDON, A., 1996, The 3D Marching Lines Algorithm. *Graphical Models and Image Processing*, 58(6), p. 503-509. Electronic version: http://www.inria.fr/RRRT/RR-1881.html

THIRION, J.P., SUBSOL, G. & DEAN, D., 1996, Cross Validation of Three Inter-Patients Matching Methods. In K.H. HÖHNE, R. KIKINIS, eds., *Visualization in Biomedical Computing*, vol. 1131, Lecture Notes in Computer Science, p. 327-336, Hamburg (Germany), Springer.

THIRION, J.P., PRIMA, S., SUBSOL, G. & ROBERTS, N., 2000, Statistical Analysis of Normal and Abnormal Dissymmetry in Volumetric Medical Images. *Medical Image Analysis*, 4 p. 111-121.

THOMPSON, D'A.W., 1917, *On Growth and Form.* Bonner, J.T. ed,. Cambridge University Press (first edition in 1917).

THOMPSON, J.L. & ILLERHAUS, B., 1998, A new reconstruction of the Le Moustier 1 skull and investigation of internal structures using 3-D-iCT data. *Journal of Human Evolution*, 35, p. 645-667.

WEBER, G.W., 2001, Virtual Anthropology (VA): A Call for Glasnost in Paleoanthropology. *The Anatomical Record (New Anat.)*, 265:193-201.

ZOLLIKOFER, C.P.E., PONCE DE LEÓN, M.S. & MARTIN, R.D., 1998, Computer-Assisted Paleoanthropology. *Evolutionary Anthropology*, 6, p. 41-54.

VIRTUAL PALEOANTHROPOLOGY: THE 4TH DIMENSION

Christoph P. E. ZOLLIKOFER & Marcia S. PONCE DE LEÓN

Résumé: La paléoanthropologie utilise des outils informatiques pour l'acquisition tridimensionnelle de données numériques, la reconstitution virtuelle de fossiles fragmentaires et la morphométrie assistée par ordinateur. Néanmoins, il faut considérer qu'au cours de la formation de fossiles la dimension temporelle est aussi importante que les trois dimensions spatiales. Le facteur temps, la «quatrième dimension», intervient dans trois composantes différentes mais interdépendantes en paléoanthropologie: l'ontogenèse (le développement individuel), la phylogenèse (la spéciation) et la diagenèse (la fossilisation). – Nous proposons des méthodes morphométriques assistées par ordinateur qui permettent d'analyser les effets du temps sur la morphologie des fossiles selon ces trois échelles.

Abstract: In paleoanthropology, computer-based technologies are used as tools for 3-dimensional data acquisition, virtual fossil reconstruction and virtual morphometry. Nevertheless, one must take account of the fact that during the formation of fossils the temporal dimension is equally important as the spatial dimensions. Time – the fourth dimension – has three distinct yet interconnected aspects in paleoanthropology, namely ontogeny, phylogeny and diagenesis. – We propose computer-based models of geometric morphometric analysis that deal with the effects of time on fossil morphology at these three levels.

INTRODUCTION

One of the major difficulties that arises during the analysis of fossil hominid morphologies is the scarcity and incompleteness of the fossil remains. Their reconstruction as well as their morphometric analysis are typically considered to represent essentially 3-dimensional tasks, in which «static» organismic structures are compared with each other and with extant samples. In this paper we investigate the role of time as the fourth dimension of virtual fossil reconstruction and morphometry. In evolutionary biology – especially in paleontology – we may discriminate between three different but interconnected aspects of temporal processes that affect the morphology of an organism: ontogeny, phylogeny and diagenesis. Accordingly, it is sensible to discriminate between ontogenetic, phylogenetic and diagenetic time scales, along which fossils must be analyzed. Morphogenesis accumulates form change over ontogenetic time, i.e. an individual's development and, in a wider sense, its entire lifetime. This process is constrained by the genome, which itself accumulates and integrates ontogenetic changes over phylogenetic time scales. While the morphology of an extant organism represents the result of processes on phylogenetic and ontogenetic time scales, diagenesis as a post-mortem process represents an additional phase of morphological modification. It is evident that the loss of information during this latter process is dramatic, both in terms of the number of available specimens and – within specimens – in terms of morphology and anatomy.

From this perspective, fossils can be characterized as organisms encountered at some (3-dimensional) point in diagenetic, phylogenetic and ontogenetic time. Accordingly, their reconstruction represents the task of resetting the diagenetic clock to zero, and their comparative morphometric analysis corresponds to the task of inferring their position along the ontogenetic and phylogenetic time scales. In practice, this endeavor is complex, since the three time scales are intertwined along one single physical dimension of time and none of them can be observed directly in fossils. Here, we report on concepts and methods for the analysis of fossil morphologies in 4 dimensions and the disentanglement of the three time scales.

RESETTING THE DIAGENETIC CLOCK: COMPUTER-ASSISTED FOSSIL RECONSTRUCTION

Over the past few years, we developed a suite of computer-assisted methods, which permit the acquisition of volume data from fossil hominids, their electronic preparation on a computer screen, the recomposition of isolated fragments in virtual reality, the morphometric analysis of virtual fossils and their computer-guided physical replication. Computer-assisted paleoanthropology (CAP) combines techniques of computer tomography (CT), computer graphics and rapid prototyping (stereolithography), which permit a completely non-invasive approach to the study of fossils (Zollikofer *et al.*, 1995; Seidler *et al.*, 1997; Spoor & Zonneveld, 1997; Conroy *et al.*, 1998; Zollikofer *et al.*, 1998).

How can the diagenetic clock be reset with these methods? The principal aim of this task is to recover the original state of the specimen's skeletal structure at the time of its death. To do so, specimens must first be freed from matrix, then restored through the re-fitting of scattered and isolated fragments and finally brought into their original shape through the correction of diagenetic deformation. The former two steps are essentially spatial tasks; they exploit topological relationships between various anatomical structures preserved in the isolated fossil fragments and rely on the general bilateral symmetry of the vertebrate skeleton (Zollikofer *et al.*, 1998;

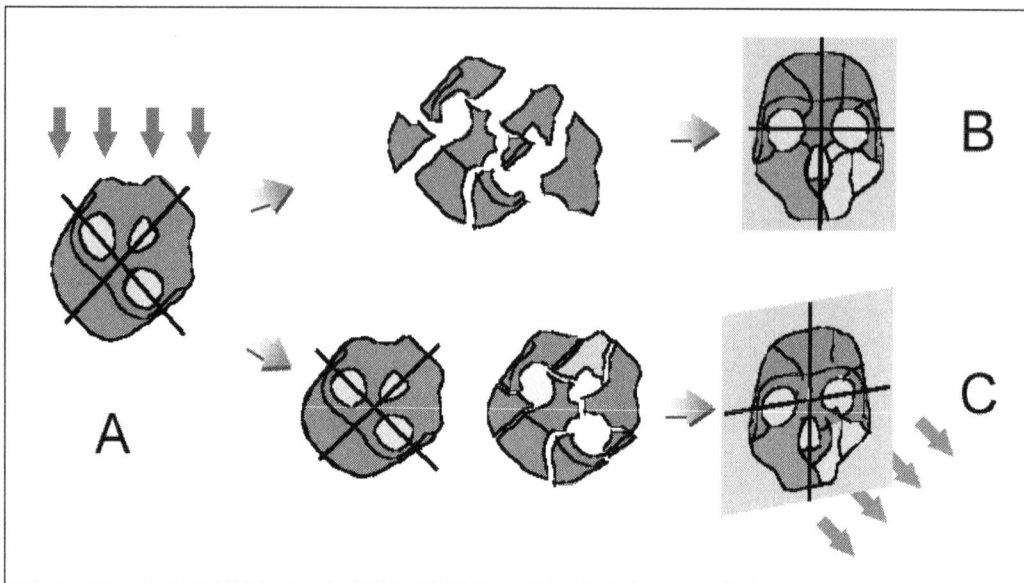

Figure 1 - Taphonomic deformation and its virtual correction. Loading exerted by overlaying strata (**A**, arrows) may result in fracturing (**B**) or plastic deformation (**C**) of fossilizing bone. In (**B**) the reassembly of the dislocated fragments results in only little residual deformation. In (**C**), residual deformation in the recomposed fossil manifests itself by systematic deviation from bilateral symmetry, which can be corrected by reversal of the deformation along the principal direction of compression (arrows).

Ponce de León & Zollikofer, 1999), such that missing parts on one side of the morphology can tentatively be completed with mirror-images of preserved counterparts. To correct taphonomic deformation, both ontogenetic and diagenetic time scales must be considered, as the deformation observed in a fossil may represent the combined effects of in-vivo and post-mortem processes. First and foremost, therefore, any potential deviation from standard anatomical conditions must be ckecked against normal anatomical variation and potential cultural modification, and assessed using paleopathological criteria. Once ontogenetic (*in-vivo*) effects have been identified, the residual deformation can tentatively be attributed to diagenetic (*post-mortem*) processes.

The original *in-situ* position and orientation of the fossil fragments relative to each other and relative to the embedding strata as well as the geometry of the deformed fossil convey important information about the spatial and temporal nature of these processes. From a biomechanical perspective, bone embedded in sediments can yield to loading by overlaying sediments in two different, but not mutually exclusive, ways (Fig. 1): (a) it may behave as a rigid material and fracture along lines of maximum strain, while the resulting fragments are dislocated relative to each other; (b) it may behave as a ductile material and undergo plastic deformation, with relatively little dislocation of neighboring anatomical structures. As was shown in a theoretical model (Ponce de León & Zollikofer, 1999), under scenario (a), the recovered morphology remains relatively undisturbed with respect to bilateral symmetry, while under scenario (b), the fossil exhibits substantial deviation from symmetry. Accordingly, the effects of fracturing deformation (a) are typically already corrected during fossil reconstruction, while plastic deformation (b) manifests itself most conspicuously by left/right discordance, which can be assessed and measured in virtual reality by

superposition of parts from one side with their mirror-imaged counterparts. Based on this information the temporal sequence of diagenetic events that led to the observed *post-mortem* modifications can be inferred, and deformation can be corrected as follows: As the outcome of plastic deformation depends on the direction of gravitational forces relative to the anatomy (Fig. 1), the virtual specimen must be positioned «*in situ*» on the computer screen. Subsequently, the taphonomic forces are simulated in reverse spatial and temporal order to recover the original morphology. It is evident that these procedures remain tentative, as the disentanglement of ontogenetic from diagenetic modifications may be incomplete, and as the complex succession of diagenetic events cannot be fully resolved in every instance.

VISUALIZING FORM CHANGE DURING ONTOGENY AND PHYLOGENY

In traditional paleoanthropology, the form of specimens is characterized by a suite of linear and angular measurements taken between anatomical points of reference, which are chosen to represent locations of biological homology. Accordingly, form change and/or difference is studied with multivariate analysis applied to arrays of such measurements. Linear and angular data are readily gathered «by hand» (once access to fossil specimens is granted), but a major disadvantage of their analysis arises from the fact that the original geometric interrelations between the *positions* of the landmarks in space are only partially represented. Over the past few years, geometric morphometric methods (GMM) have opened up new ways in which variation of organismic form can be measured and treated statistically (Rohlf & Marcus, 1993). These methods use configurations of anatomical landmarks to characterize the form of specimens. Evaluating the 2- or 3-dimensional position of landmarks is technically more demanding than the measurement of interlandmark distances and typically involves digitizing devices or – in a CAP environment – interactive measuring tools permitting 3-dimensional data sampling from virtual fossil specimens.

Landmark configurations simultaneously represent locations of biological *and* geometric homology, such that the real-space geometric properties of the biological structures under investigation can be fully integrated into the subequent multivariate analyses. The mathematical equivalent of the biological homology relation can be established in various ways. A first possibility consists in the exhaustive analysis of all between-landmark distances in a configuration (Euclidean Distance Matrix Analysis, EDMA; Lele, 1993; Richtsmeier & Lele, 1993). The second possibility is to superimpose the specimen's landmark configurations according to an optimum criterion (e.g. Procrustes superimposition by minimizing the sum of squared distances from an average configuration; Rohlf & Slice, 1990). This procedure eliminates differences in position, orientation and scaling (size) between the configurations and establishes a «linearized Procrustes shape space» (Rohlf & Slice, 1990), in which the shape of each specimen is given by its linear deviation (landmark coordinate by landmark coordinate) from the average configuration, the so-called consensus). These linearized «Procrustes residuals» can then be analyzed with classical methods of multivariate analysis, e.g. using principal components analysis (PCA) to reduce the dimensionality of the analysis and to capture major trends of shape change in the sample. The third possibility – which may be combined with the second – is especially appealing for the visualization of shape change and shape difference. It uses Thin Plate Spline interpolants (TPS), which act as spatially pervasive non-linear deformation functions between landmark configurations (Bookstein, 1991). PCA can be performed on the TPS coefficients characterizing each specimen's deviation from a consensus configuration, such that major trends of shape change in the sample can directly be expressed and visualized as deformation functions applied to the consensus (Bookstein, 1991). To visualize complex patterns of shape change and/or difference between specimens, D'Arcy W. Thompson introduced the ingenious concept of deformation grids, which provide an immediate visual grasp of the transformation of one into another morphology (Thompson, 1948). The TPS functions introduced by Bookstein (1991) provide an elegant mathematical formulation and extension of the graphical concepts of Thompson.

GMM are well-suited to study shape variation in fossil morphologies, notably of the cranium, since the skull is a rigid structure, whose surface displays a large number of landmarks representing locations of homology. Notably TPS analysis and visualization is being used extensively to study cranial shape change in various vertebrate species (Zelditch *et al.*, 1992; Loy *et al.*, 1993; Lynch *et al.*, 1996; Yaroch, 1996; O'Higgins & Jones, 1998; Rao & Suryawanshi, 1998; O'Higgins, 2000). Most studies, however, confine themselves to the analysis of 2-dimensional data sets – typically

projections of landmark coordinates onto the midsagittal plane or data from one side of the skull. In fact, during the visualization of patterns of shape variability in 3 dimensions, the utility of Thompson-style deformation grids is limited for several reasons. The first is a practical one. 3D cuboid grids or 2D square grids positioned in space (and projected onto paper) tend to be unintelligible, since our visual attention is directed towards undesired boundary effects at the edges of the grid (Dryden & Mardia, 1998; O'Higgins & Jones, 1998). The second reason is a biological one: Deformation grids introduce an external system of geometric reference which is not directly related to the original biological structure. This makes it difficult to interpret geometric «deformation» in terms of potential biological mechanisms of shape variation, both in time and space.

A possible solution to this problem consists in taking account of the fact that the original landmarks used for the geometric morphometric analyses are typically derived from surface structures (Zollikofer & Ponce de León, in press). During visualization, it is therefore sensible to make explicit reference to the surfaces of the specimens rather than to an arbitrary Cartesian coordinate grid. To do so, it is necessary to re-express the TPS transformation functions resulting from geometric morphometric analysis as deformations of a specimen's object surface. In engineering terms, a TPS function can be imagined as a displacement field: for every point in space, an individual vector of displacement defines the local amount and direction of shape change. Accordingly, for the visualization of the specimen's shape change, we only consider displacement vectors at object surfaces. With this procedure, the original problem of visualizing volume deformation is reduced to the visualization of 2-dimensional surface deformation in 3D space. However, 3D surface vectors of shape change depicted on paper (i.e. 2D) are still difficult to grasp visually. It is therefore sensible to decompose the displacement vectors into two components, one vertical and one parallel to the local surface orientation, respectively (Fig. 2). This permits the visualization of patterns of shape change in directions normal and tangential to the object surface. While the normal component of shape change can be visualized using color- or gray-scale coding, the tangential component can be rendered as a vector field spread out over the cranial surface (see Fig. 4).

The biological motivation of this procedure is as follows: Organs typically grow through cell divisions *at tissue surfaces*. Accordingly, skeletal structures grow through apposition/ resorption of extracellular matrix perpendicular to the bone surface. During development of complex 3-dimensional structures such as the vertebrate skull, anatomical subregions grow by surface-bounded expansion or drift, and/or by passive displacement relative to neighboring structures (Moss & Young, 1960; Enlow, 1990). The independent visualization of the normal and tangential components of shape change thus provides a tentative visual grasp of how much and in which direction anatomical subregions expand, drift and/or are displaced relative to each other.

How does this method of visualization impinge on the geometric morphometric analysis of *temporal* changes in

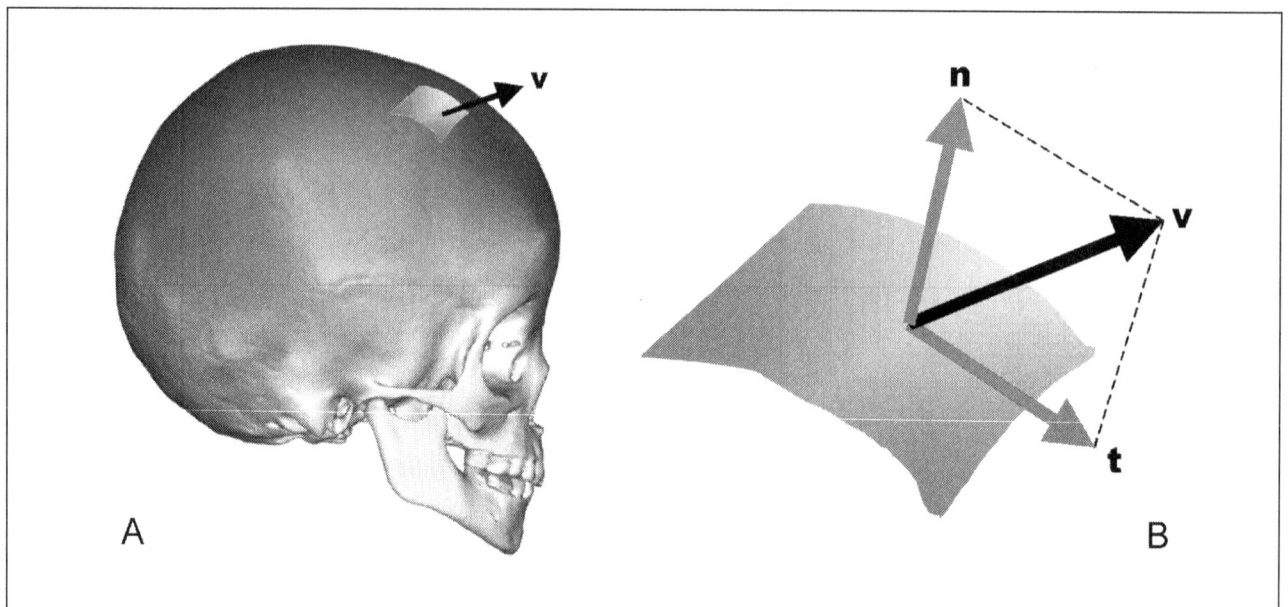

Figure 2 - Decomposition of a 3-dimensional displacement vector relative to the surface of a cranium. **A**: a small cranial surface patch and its associated displacement vector **v**. **B**: Decomposition of v into its normal (**n**) and tangential (**t**) components.

fossil morphologies? Any *patterns* of shape change (e.g. along an ontogenetic trajectory) or shape difference (e.g. between specimens belonging to different species) revealed with GMM ultimately reflect different stages or different types of ontogenetic *processes*. The proposed method of visualization greatly facilitates the interpretation of the observed patterns in terms of underlying processes. As one example, we consider here the comparative geometric morphometric analysis of growth and development in Neanderthals and modern humans (Ponce de León & Zollikofer, 2001). Using methods of CAP, an ontogenetic series of Neanderthal crania comprising specimens from dental age 2 through to adulthood was reconstructed, and a similar ontogenetic series of anatomically modern humans (both fossil and extant) was used for comparison. The form of each skull in the pooled Neanderthal/modern human sample was characterized by the 3D coordinates of 63 landmarks. Generalized least-squares superposition was used to evaluate the centroid size (Bookstein 1991) and linearized Procrustes shape (Rohlf & Slice, 1990) of each specimen. Relative Warp (RW) analysis (Rohlf, 1993) was used to capture major trends of shape variation in the sample. As a result, shape variation related to dental age, centroid size and taxon was concentrated in the first two relative warps (RW1 and RW2, Fig. 3), which account for 60.6% of the total shape variation in the sample. The Neanderthal and modern human subsamples are separated from each other along RW2 (10.1% of the total shape variation), while the specimens in each subsample are ordered along RW1 according to their age (50.5% of the total shape variation). One salient feature of this analysis is that within-species shape variation (i.e. along ontogenetic trajectories) is considerably larger than shape variation between species (i.e. across ontogenetic trajectories). Nevertheless, the characteristic difference in shape between Neanderthals and modern humans is already established around the age of 2

years and remains unchanged during further development, as evinced by the fully separated, parallel postnatal ontogenetic trajectories of the two species.

Overall, the utilization of GMM permits the characterization of Neanderthals and modern humans as separate dynamical «ontogenetic entities» rather than static «morphological structures». This difference in perspective is crucial, since it impinges on the question as to whether Neanderthals and modern humans must be considered separate species or subspecific variants of the same species. Adopting the static, structural perspective (e.g. by restricting the comparison to adults or peer groups of both taxa,) tends to overemphasize within-taxon relative to between-taxon variation, since the developmental background of variation is not considered. The dynamical approach, on the other hand, reveals that taxon-specific differences appear early and persist during later development, despite considerable within-taxon variation along the ontogenetic trajectory. The parallel trajectories further indicate that, once the taxon-specific differences are established (around age 2) the development of Neanderthals and modern humans is largely similar. One additional ontogenetic inference that can be made on the basis on these data is that the developmental processes that account for the principal differences between Neanderthals and modern humans must have occurred early during development. This early differentiation further hints at a strong genetic rather than environmental background of the morphological differences between these taxa.

The proposed graphical methods can now be used to visualize the morphological effects of (a) shape change corresponding to the advancement along the parallel ontogenetic trajectories (arrow along RW1 in Fig. 3) and (b) shape difference

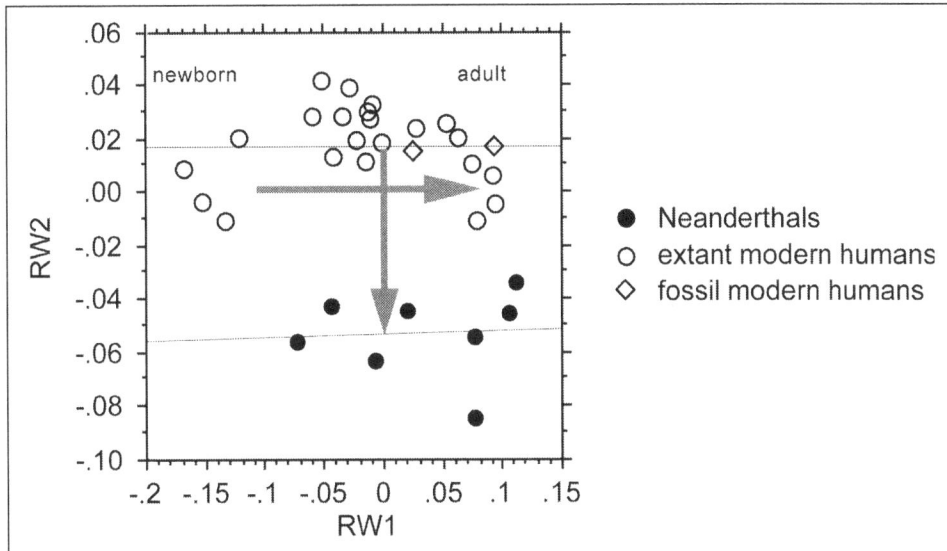

Figure 3 - Relative Warp analysis of an ontogenetic series of Neanderthal specimens (from left to right: Pech de l'Azé, Roc de Marsal, Devil's Tower, Teshik, Tash, Le Moustier, Tabun 1, Forbes' Quarry, La Ferrassie 1) and modern human craniomandibular specimens (fossil specimens: Qafzeh 9, Qafzeh 11). Taxa are clearly separated along relative warp 2 (RW2), while each of them follows an ontogenetic trajectory parallel to relative warp 1 (RW1).

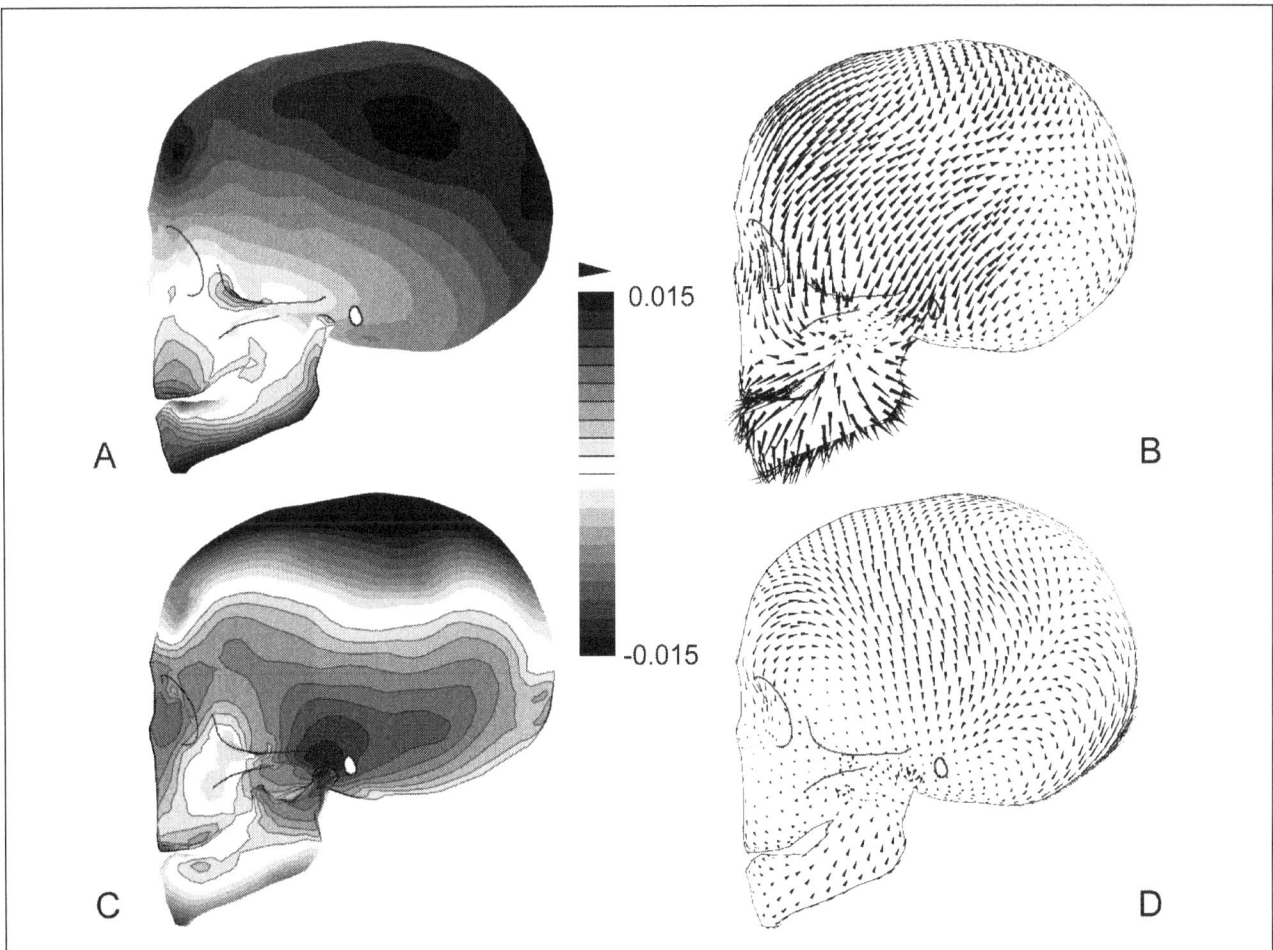

Figure 4 - Visualization of Neanderthal/modern human shared craniomandibular growth patterns (**A**, **B**, corresponding to advancement along horizontal arrow in Fig. 3) and patterns of shape differences between Neanderthals and modern humans (**C**, **D**, corresponding to vertical arrow in Fig. 3). **A,C/B,D**: normal/tangential vector maps accounting for shape difference perpendicular/parallel to the cranial surface (in **A,C**, shaded and outlined-shaded areas indicate inward and outward orientation of the normal vectors, respectively; scale is in units of centroid size). Growth patterns are characterized by expansion of the viscerocranial parts, as opposed to relative contraction of the neurocranium. Neanderthals, relative to modern humans, show broadening of the temporomandibular region, flattening of the cranial vault and projection of the face and occipital region.

corresponding to the distance between the trajectories (arrow along RW2 in Fig. 3). The resulting patterns of shape change/ shape difference are visualized in Fig. 4. As indicated by the parallel trajectories along RW1, patterns of morphological change during postnatal development are similar in Neanderthals and modern humans and can therefore be visualized using a Neanderthal/modern human consensus cranium (Figs. 4 B). The most conspicuous feature arising from these graphs is the distinct pattern of development of the viscerocranium and the neurocranium: while the former expands in anterior and caudal direction, the latter exhibits relative «contraction», i.e. it assumes a steadily smaller proportion of the growing skull. This pattern most probably represents a general developmental feature of the hominid cranium.

The difference in craniomandibular shape between Neanderthals and modern humans (Figs. 4 C, D) is visualized as the pattern of shape difference resulting from transformation of a modern human into a Neanderthal, using the Neanderthal/modern human consensus cranium. Note that this pattern of shape transformation expresses the distance between the taxon-specific ontogenetic trajectories through shape space and is therefore independent of the age at which the comparison is performed (the consensus cranium represents an juvenile individual of approx. 8 years of age). Figs. 4 C, D therefore account for the principal morphological features of the cranium of Neanderthals relative to that of modern humans. The graphs show a suite of classical Neanderthal features such as a relatively flat cranial vault with a broad temporal region, projection of the midface face and occipital region, and a receding symphyseal region of the mandible (Stringer & Gamble, 1993). Since all these features are already established around the age of 2 years, it can readily be inferred that the taxon-specific Neanderthal and modern human morphologies were essentially brought about prior to that age, probably already during prenatal development (Lieberman, 1998; Lieberman, 2000; Ponce de León & Zollikofer, 2001). From this perspective, the «static» postnatal differences between taxa represent contrasts between taxon-specific early developmental programs.

Combining these observations, the following conclusions can be drawn: Neanderthals and modern humans share their postnatal patterns (most probably also their processes) of development, as evinced by their parallel trajectories through shape space. The phyletic difference in shape between these species, as quantified and visualized by the distance between their postnatal ontogenetic trajectories, reflects evolutionary modifications leading to divergent prenatal/early postnatal developmental programs. Fig. 3A therefore encapsulates shape change on both ontogenetic and phylogenetic time scales, corresponding to RW1 and RW2, respectively. The pattern of taxon-specific difference can be linked to (no longer observable) differences in early ontogenetic process. – This example demonstrates that virtual paleoanthropology is an indispensable tool for the analysis of spatiotemporal change – i.e. change in four dimensions – at the level of diagenesis, phylogeny and ontogeny.

Acknowledgements

We would like to thank Professor Peter Stucki, director of the MultiMedia Laboratory, for constant support of our research. Lorenzo Rook's geological advice is gratefully acknowledged. This work was supported by grant #31-42419.94 of the Swiss National Science Foundation and a habilitation grant of the Canton of Zurich.

Authors' addresses:

Ch. P. E. ZOLLIKOFER (zolli@ifi.unizh.ch; http:// www.ifi.unizh.ch/staff/zolli)

M. S. PONCE DE LEON (marcia@ifi.unizh.ch; http:// www.ifi.unizh.ch/staff/marcia)

Anthropologisches Institut

und MultiMedia Laboratorium/Institut für Informatik

Universität Zürich-Irchel, CH-8057 Zürich/Schweiz

BIBLIOGRAPHY

BOOKSTEIN, F. L., 1991, *Morphometric Tools for Landmark Data*. Cambridge: Cambridge University Press.

CONROY, G., WEBER, G., SEIDLER, H., TOBIAS, P., KANE, A. & BRUNSDEN, B., 1998, Endocranial capacity in an early hominid cranium from Sterkfontein, South Africa. *Science* 280, 1730-1.

DRYDEN, I. L. & MARDIA, K., 1998, *Statistical Shape Analysis*. New York: John Wiley & Sons.

ENLOW, D. H., 1990, *Facial Growth*. Philadelphia: Saunders.

LELE, S., 1993, Euclidean distance matrix analysis (EDMA): estimation of mean form and mean form difference. *Math. Geol.* 25, 573-602.

LIEBERMAN, D. E., 1998, Sphenoid shortening and the evolution of modern human cranial shape. *Nature* 393, 158-162.

LIEBERMAN, D. E., 2000, Ontogeny, homology and phylogeny in the hominid craniofacial skeleton: the problem of the browridge. In *Development, Growth and Evolution*, edited by P. O'Higgins & M. J. Cohn. London: The Linnean Society of London, p. 85-122.

LOY, A., CORTI, M. & MARCUS, L. F., 1993, Landmark data: size and shape analysis in systematics. A case study on Old World Talpidae (Mammalia, Insectivora). In *Contributions to Morphometrics*, edited by L. Marcus, E. Bello & A. García-Valdecasas. Madrid: Consejo superior de investigaciones científicas, p. 215-240.

LYNCH, J. M., WOOD, C. G. & LUBOGA, S. A., 1996, Geometric morphometrics in primatology: craniofacial variation in *Homo sapiens* and *Pan troglodytes*. *Folia Primatol.* 67, 15-39.

MOSS, M. L. & YOUNG, R. W., 1960, A functional approach to craniology. *Am. J. Phys. Anthropol.* 18, 281-292.

O'HIGGINS, P., 2000, The study of morphological variation in the hominid fossil record: biology, landmarks and geometry. *J. Anat.* 197, 103-120.

O'HIGGINS, P. & JONES, N., 1998, Facial growth in *Cercocebus torquatus*: an application of three-dimensional geometric morphometric techniques to the study of morphological variation. *J. Anat.* 193, 251-272.

PONCE DE LEÓN, M. S. & ZOLLIKOFER, C. P. E., 1999, New evidence from Le Moustier 1: Computer-assisted reconstruction and morphometry of the skull. *Anat. Rec.* 254, 474-489.

PONCE DE LEÓN, M. S. & ZOLLIKOFER, C. P. E., 2001, Neanderthal cranial ontogeny and its implications for late hominid diversity. *Nature* 412, 534-538.

RAO, C. R. & SURYAWANSHI, S., 1998, Statistical analysis of shape through triangulation of landmarks: a study of sexual dimorphism in hominids. *Proc. Natl. Acad. Sci. U.S.A.* 95, 4121-4125.

RICHTSMEIER, J. T. & LELE, S., 1993, A coordinate-free approach to the analysis of growth patterns: models and theoretical considerations. *Biol. Rev.* 68, 381-411.

ROHLF, F. J., 1993, Relative warp analysis and an example of its application to mosquito wings. In *Contributions to Morphometrics*, edited by L. Marcus, E. Bello & A. García-Valdecasas. Madrid: Consejo superior de investigaciones científicas, p. 131-159.

ROHLF, F. J. & MARCUS, L., 1993, A revolution in morphometrics. *Trends Ecol. Evol.* 8, 129-132.

ROHLF, F. J. & SLICE, D., 1990, Extensions of the Procrustes method for the optimal superimposition of landmarks. *Syst. Zool.* 39, 40-59.

SEIDLER, H., FALK, D., STRINGER, C., WILFING, H., MULLER, G. B., ZUR NEDDEN, D., WEBER, G. W., REICHEIS, W. & ARSUAGA, J., 1997, A comparative study of stereolithographically modelled skulls of Petralona and Broken Hill: implications for future studies of middle Pleistocene hominid evolution. *J. Hum. Evol.* 33, 691-703.

SPOOR, F. & ZONNEVELD, F., 1997, CT-based 3-D imaging of hominid fossils, with notes on internal features of the Broken Hill 1 and SK 47 crania. In *The paranasal sinuses of higher primates: development, function and evolution*, edited by T. KOPPE & H. NAGAI. Tokyo: University of Tokyo Press.

STRINGER, C. B. & GAMBLE, C., 1993, *In Search of the Neanderthals: Solving the Puzzle of Human Origins*. London: Thames and Hudson.

THOMPSON, D'A. W., 1948, *On Growth and Form*. Cambridge: Cambridge University Press.

YAROCH, L. A., 1996, Shape analysis using the thin-plate spline: Neanderthal cranial shape as an example. *Yb. Phys. Anthropol.* 39, 43-89.

ZELDITCH, M. L., BOOKSTEIN, F. L. & LUNDRIGAN, B. L., 1992, Ontogeny of integrated skull growth in the cotton rat *Sigmodon fulviventer*. *Evolution* 45, 1154-1180.

ZOLLIKOFER, C. P. E. & PONCE DE LEÓN, M. S., in press, Visualizing patterns of craniofacial shape variation in *Homo sapiens*. *Proc. Roy. Soc. B* .

ZOLLIKOFER, C. P. E., PONCE DE LEÓN, M. S. & MARTIN, R. D., 1998, Computer-assisted paleoanthropology. *Evol. Anthropol.* 6, 41-54.

ZOLLIKOFER, C. P. E., PONCE DE LEÓN, M. S., MARTIN, R. D. & STUCKI, P., 1995, Neanderthal computer skulls. *Nature* 375, 283-285.

3-D IMAGING AND TRADITIONAL MORPHOMETRIC ANALYSIS OF THE ADOLESCENT NEANDERTAL FROM LE MOUSTIER

Jennifer L. THOMPSON, Andrew J. NELSON, Bernhard ILLERHAUS

Résumé: Les auteurs se proposent de montrer l'étendue des informations qui peuvent être utilisées par les paléoanthropologues pour décrire les hominidés fossiles, analyser leurs variations de dimensions et de proportions et interpréter ces données grâce aux nouvelles techniques d'imageries tridimensionnelles associées aux méthodes d'analyses plus conventionnelles. En prenant comme exemple le crâne de l'adolescent néandertalien Le Moustier I, les auteurs montrent l'intérêt de l'association de ces deux méthodologies pour l'étude et la description du crâne, l'étude de la denture et du squelette post-crânien de ce fossile. Les résultats obtenues sont ensuite remplacés dans une plus large perspective en démontrant leurs intérêt pour l'étude de la croissance et du développement et l'analyse phylogénique.

Abstract: The purpose of this paper is to outline information that can be used by paleoanthropologists to describe fossil hominid morphology, to document size and shape variables, as well as to interpret those data within a broader context using 3-D imaging techniques as well as traditional analytical methods. Using the Le Moustier 1 adolescent Neandertal as an example, this paper will demonstrate the benefits of both types of techniques for the analysis of fossil hominid specimens. It will outline how these techniques were used to describe the skull, dentition, and postcranial skeleton, to identify species specific features, and to determine the age and sex of the Le Moustier 1 specimen. It will then place these data into a broader analytical context to address both growth & development issues and phylogenetic implications of the findings.

INTRODUCTION

With the discovery of any new fossil hominid specimen, the task of the paleoanthropologist is to undertake a description of the specimen and its state of preservation, describe the specimen's anatomy, compare that anatomy to other fossils of similar antiquity, to determine shared species characteristics, the chronological age of the specimen, its sex, and any unique features of the individual. This information is then used to place the individual into a broader phylogenetic, and, if appropriate, ontogenetic context. To achieve these goals, paleoanthropologists have, until recently, relied solely on traditional morphometric measurements to measure size and shape of skeletal elements, including the skull, teeth, and postcrania. They have also relied on non-metric traits to examine features such as muscle development or the presence/ absence of shared, shared derived, or derived, species specific characteristics. While these techniques have produced a rich body of research, paleoanatomists' efforts have been hampered in two ways. First of all, not all fossils are complete. Second, valuable information about internal morphology of skeletal elements is often missed, due to the difficulty of undertaking radiographic analysis and/or the difficulty of interpreting two-dimensional (2-D) plane film radiographs.

The field of paleoanthropology has been revolutionized by the addition of computed, or computerized tomography to our battery of analytical tools. Computerized tomography (CT) images have contributed to our knowledge of the internal morphology and to our ability to study overall dimensions of incomplete or fragmentary fossil skulls (Tate & Cann 1982; Conroy & Vannier 1984, 1986; Wind 1984; Zonneveld & Wind 1985; Zonneveld et al. 1989; Conroy et al. 1990; Spoor et al. 1993; Spoor & Zonneveld 1994, 1995; Spoor et al. 1994; Hublin et al. 1996; Thompson & Illerhaus 1998, 2000, in press). This technique can be used to produce 2-D "slices" through a fossil, or to produce three-dimensional (3-D) images. The latter ability allows one to make virtual reconstructions of skulls from their various fragments without damaging the original specimen (Kalvin et al. 1992; Thompson & Illerhaus 1998, 2000, in press; Zollikofer et al. 1995, 1998). CT imaging has also been used to investigate internal structures of the skull such as the labyrinth of the inner ear and the frontal sinus as well as the dentition of fossil hominids (e.g. Spoor & Zonneveld 1995: Spoor et al. 1993; Hublin et al. 1996; Seidler et al. 1997; Thompson & Illerhaus 1998, 2000, in press).

Given the focus of this symposium it seemed appropriate to discuss the application of 3-D imaging in paleoanthropology by demonstrating the use of both traditional morphometric methods as well as CT methods using Le Moustier 1 as a case study. This individual is an adolescent Neandertal, discovered in 1908 at the site of Le Moustier in the Dordogne of France (Hauser 1909). It was described briefly when first discovered (Klaatsch 1909a, b, c, d; Klaatsch & Hauser 1909) but, by comparison to today's standards was, until recently, relatively unstudied (Bilsborough & Thompson in press a). Work on this specimen began again in 1992 and, since then one of us (JT) along with various co-authors undertook a description and analysis of the skull, dentition and postcranial skeleton, as well as a CT or virtual reconstruction of the skull. The intent of this paper is to summarize some of these results, highlighting the contribution of both traditional and CT methods to our interpretation of the morphology of this specimen.

DESCRIPTION OF SPECIMEN & SPECIES SPECIFIC CHARACTERISTICS - SKULL

Weinert's (1925) original reconstruction of the Le Moustier 1 cranium was dismantled during the late 1960's and Figure 1 illustrates the current disassembled state of the skull and dentition. With Alan Bilsborough of Durham University, UK, JT undertook a detailed anatomical description of the specimen using traditional methods (Thompson & Bilsborough 1997, 1998a, b, in press; Thompson in press). In these papers, we used conventional morphometric techniques to describe typical Neandertal features such as cranial size and shape (long, low skull), the state of particular anatomical traits (e.g. occipital bun, iniac fossa, mastoid tubercle, shape of the orbits) and so on. We noted that some Neandertal features are not fully developed, due to the immaturity of the specimen. These features include the lack of a retromolar space, the gracile supraorbital torus, and the lack of mid-facial prognathism. We also identified unique features such as the asymmetry of the entoglenoid processes of the cranial base and the misshapen left mandibular condyle – likely broken and healed while the individual was still living.

Early in 1994, JT and BI were invited to undertake a virtual CT reconstruction of the skull by the director of the Museum fur Vor- und Fruhgeschicte, Berlin where the specimen is housed. This provided us with the opportunity to further our analysis of the morphology of the specimen (Hoffmann 1997; Hoffmann *et al.* 1994). A full description of the reconstruction process, using microcomputed tomography data, is outlined elsewhere (Thompson & Illerhaus 1998). The microcomputed tomography data were collected for this analysis using the 3-D micro-tomograph housed at Bundesanstalt für Materialforschung und -Prüfung (BAM), Berlin using methods developed at BAM (e.g. Illerhaus *et al.* 1994; Illerhaus *et al.* 1997 a, b).

Figure 2 illustrates several views of the virtual reconstruction of the Le Moustier 1 skull. This figure illustrates that the overall shape of the skull is typical of adult Neandertals — the long, low skull, and occipital bun are evident, as is the youthful features mentioned above – the lack of retromolar space, undeveloped mid-facial prognathism, etc.

Aside from the reconstruction, two additional objectives were successfully achieved using the CT data: 1) we were able to differentiate between real fossilized bone and reconstructed material (plaster etc.) and 2) we were able to examine internal structures of the skull. Analysis revealed that much of the palate was composed of reconstructed material, as were several cranial pieces (see Thompson & Illerhaus 1998 for details). Access to the CT data allowed us to examine the frontal sinuses in virtual space. Neandertal sinuses usually spread to mid-orbit but do not extend up onto the frontal bone (Heim 1978; Tillier 1974; Trinkaus 1983; Vlcek 1967, 1969). This individual's sinuses are not likely to have been fully grown, explaining why they do not extend to mid-orbit. CT data also allowed us to isolate the labyrinth of the inner ear. Analysis reveals that Le Moustier 1 follows the Neanderthal pattern of having an inferiorly positioned posterior semicircular canal relative to the plane of the lateral canal (Thompson & Illerhaus 1998, 2000, in press).

DESCRIPTION OF SPECIMEN & SPECIES SPECIFIC CHARACTERISTICS - DENTITION

A detailed description and analysis of the dentition has also been undertaken using traditional methods (Bilsborough & Thompson in press b). In that paper, we discussed the preservation of the dentition, development of marginal ridges, accessory cusps, dental wear, dental metrics. The paper also

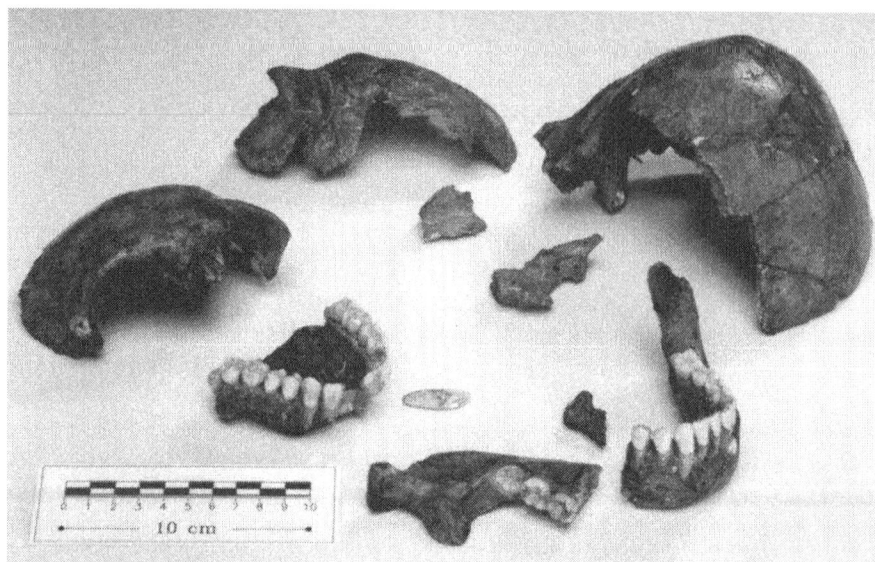

Figure 1 - Le Moustier I skull and dentition
Image reproduced with permission from Thompson & Illerhaus,1998, *J Hum Evol* 35, 647-665.

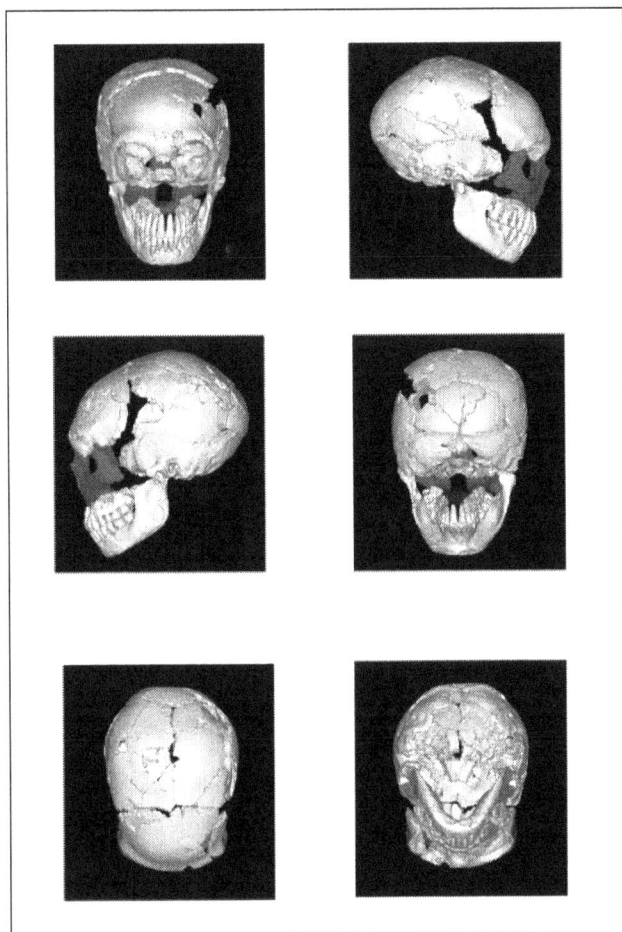

Figure 2 - Virtual reconstruction of the Le Moustier 1 skull Images reproduced with permission from Thompson & Illerhaus, 1998, *J Hum Evol* 35, 647-665.

examined features like the extent of shoveling on the upper central incisors, the presence of deep lingual pits on the upper lateral incisors, and the presence of extra cusps (e.g. Carabelli's cusp on the upper first molars and a tuberculum intermedium or metaconulid on the lower third molars). The extent of dental wear was documented. For example, the lower incisors show labial wear indicating an overbite, there is asymmetrical wear when comparing left and right dental arcades, and there is evidence of dental probing between the teeth. The presence and degree of enamel hypoplasia was also noted.

Thompson and Illerhaus (in press) used CT data to enhance the description of gross morphology by allowing us to "slice" the teeth in several planes. This analysis revealed that the upper premolars both have double roots. In the genus *Homo* it is typical for the first upper premolar to have two roots, but the second premolar to have a single root (Hillson 1996). The second upper premolars of Le Moustier 1 are fully two-rooted teeth with a larger buccal and smaller lingual root which expands our knowledge of the variation and/or presence of this trait in Neandertals. We were also able to document at the extent of taurodontism (the expanded pulp cavities of the specimen), a trait characteristic of Neandertals. We were also able to fully visualize an impacted canine which had fully formed, but never erupted. Finally, the CT analysis of the dentition played an important role in the determination of the age of this individual (see below).

DESCRIPTION OF SPECIMEN & SPECIES SPECIFIC CHARACTERISTICS - POSTCRANIA

Thompson & Nelson (in press b) provide a detailed description and metric analysis of the postcranial skeleton of Le Moustier 1, using casts and what remains of the original specimen. Unfortunately, the postcranial bones were badly damaged by fire at the end of WWII. The original material that survived the fire suffered shrinkage and some distortion from the heat (Hermann, 1977), but we were still able to determine a great deal about this individual by examining the anatomy in a comparative perspective.

The Le Moustier 1 postcranial skeleton possesses a number of typical adult Neandertal features. For example, a cast of a rib fragment is thick and triangular in cross-section, but whether the rib cage was barrel-shaped as in adult Neandertals is unknown. In fact, given the youthful age of the Le Moustier 1 individual, it is unlikely that the chest had reached its adult proportions, since those are achieved at the end of the adolescent period (Tanner 1962). According to Klaatsch (1909c, d), the Le Moustier 1 clavicle is relatively long and slender as seen in adult Neanderthals though a cast indicates that it is missing its epiphyses. According to Thompson & Nelson (in press b) the chest of Le Moustier 1 is deep. The arm bones indicate that the radial tuberosity is placed medially, the radius is strongly curved, the humerus is robust with a strongly marked deltoid tuberosity typical of adult specimens (Klaatsch & Hauser 1909), there is a prominent *M. pronator quadratus* crest on the ulna, and an anterior orientation of the trochlear notch of the ulna. Despite his youth, we can see from the well developed muscle attachment areas on the arms that the capability for strong rotation and supination of the forearm was already present in this youth. This observation has important behavioural implications (see Thompson in press; Thompson *et al.* in press, for discussion of these features).

Comparisons between the limb dimensions of this individual with those of adult Neandertals, demonstrates that in several features, such as femoral head size, Le Moustier 1 was a scaled version of the adult form (Thompson & Nelson, in press b). Le Moustier 1 was also a scaled version of the adult form in having shortened distal limbs. The arm bones show a similar pattern. Like adult Neandertals, the diaphysis of the Le Moustier 1 femur is robust, cylindrical, lacks a pilaster, has thick cortical bone and a relatively narrow marrow cavity.

The postcranial elements were not examined by CT methods as they were damaged and distorted by fire, although they were examined using plane film radiography (Hermann 1977). However, in a less damaged specimen, CT analysis would certainly be of use in accurately measuring things such as thickness and distribution of cortical bone, and the distribution and orientation of trabecular bone.

SEX DETERMINATION

From the time of its discovery in 1908, Le Moustier 1 was assumed by many researchers to be male based on its size and robusticity. However, since all Neandertals are robust compared to living modern humans, this conclusion must be approached with caution. The determination of sex also has relevance to the determination of age, as females mature earlier than males. In assessing the sex of this specimen using traditional methods, Thompson & Nelson (in press a, b) examined the size of the dentition (the teeth are large compared to both early and late Neandertals) the form of the mandibular ramus flexure and the absolute and relative size of the femoral head. These features indicate that this individual was a male.

Using CT methods Thompson & Illerhaus (2000, in press) made a virtual endocast of the specimen. The cranial capacity is 1540 cm³ (± 7.7ml). Details of this work have already been published (e.g. Illerhaus & Thompson 1999). This estimate of cranial capacity of Le Moustier 1 falls within 1 standard deviation of the male adult Neandertal mean and well above the Neandertal female mean (Thompson & Nelson in press a) — again indicating that this individual was male. Thus we concur with the original assessment that this is indeed a male individual.

AGE DETERMINATION - TEETH

Early papers published photographs of some of the teeth out of their sockets, but not all details were clear. The specimen, in its current state, shows all the dentition fully erupted, except for the impacted left permanent canine and the third molars. In its current state, the distal parts of the lower third molar crowns are obscured by matrix and all 4 molars are positioned as if they were erupting. This observation is important because, in the initial publications, Hans Klaatch, an anatomist, said that all 4 wisdom teeth were totally encased within the jaws. However, subsequently, at least twice, the teeth were removed from the jaws for study. The position of the molar teeth has an impact on one's estimate of age, so this point is rather important as we will see below. If the molars are erupting, this would indicate, by human standards, an age of 16 or more. However, if the molars were unerupted, the age at death estimate would be younger, depending upon the state of development of the crown and root of each tooth.

The 3-dimensional capabilities of the CT analysis allowed us to "cut and paste" the third molars into their crypts, demonstrating that there was ample space for them within the alveolus, suggesting that they were unerupted. Furthermore, we were also able to clearly visualize the developing roots of the molars (see Ponce de Leòn et al. 2000 for an alternative approach). The roots were only approximately 25% complete, indicating a dental age of 15.5 years, using modern human standards (see Thompson & Nelson 2000, in press a, for explanation). At this stage, this individual had achieved roughly 85% of adult dental maturity.

The ability to clearly visualize the roots is quite important, as tooth formation is generally thought to yield the most accurate estimate of an individual's age, as eruption times can vary greatly

AGE DETERMINATION – SKELETON

Work by Nelson & Thompson (2000, 2002, in press) reveals that, according to the length of his femur, the Le Moustier 1 specimen is equivalent to a 10.5 year old modern human boy. An examination of the original specimen (Thompson & Nelson in press a) revealed that the rami of the ischium and pubis were not fully fused, but had begun to unite. The fact that they are not fully united in this specimen would suggest that he is 9 years of age or less, skeletally, according to modern human standards. Expressing his estimated stature as a percentage of adult height, Le Moustier 1 matches a 12 year old modern human body who has achieved 85% of his adult stature (Nelson & Thompson 2000). Recall that the dental estimate for this individual was 15.5 year of age. A typical 15 year old modern human boy would have achieved approximately 95% of his adult stature (Nelson & Thompson 1999). Thus, it is clear that age estimates based on dental development and on postcranial growth are not in accord for this individual.

GROWTH & DEVELOPMENT ISSUES

This lack of concordance between dental and skeletal age led to an investigation of growth and developmental issues from a broader perspective (Thompson & Nelson 2000). Work by Thompson & Nelson (2000) indicated that *Homo erectus* is similar to apes in having relatively fast skeletal growth relative to dental growth. Upper Paleolithic modern humans apparently followed the recent modern human pattern. Neandertals appear to have delayed postcranial growth relative to their dental growth (or to have had fast dental growth relative to postcranial growth).

Nelson & Thompson (2001, in prep.) applied the same technique to explore the rate of growth of the skull relative to the dentition. They found that in some dimensions, such as facial height, both Neandertals and Upper Paleolithic modern humans both demonstrate a delay of cranial growth relative to dental growth. While this is superficially consistent with the delay in relative growth of the postcranium, the fact that the Neandertals demonstrate both delays, while the Upper Paleolithic sample delays facial height and not linear growth is really quite striking.

Using 3-D imaging, Thompson & Illerhaus (1998, 2000) took several metric measurements of the virtual reconstruction. These dimensions were compared between Le Moustier 1 who is dentally 15.5 years of age against another Neandertal specimen (Teshik Tash) that is about 10 years of age (as well as younger specimens), since comparison with a younger individual allows us to narrow down the timing of the attainment of adult morphologies. We note that mid-facial prognathism is not fully developed in either individual and therefore Le Moustier 1 would have likely grown a more typical Neandertal face, including a retromolar space, had he growth to full maturity. Facial prognathism would likely have increased had Le Moustier 1 lived to adulthood, by lengthening of the jaws to accommodate the eruption of the third molars, but at present his upper facial height measurement differs little from that of Teshik Tash indicating that this feature likely develops late in Neandertal ontogeny.

Nelson & Thompson (2000, 2002, in press) have examined various aspects of long bone growth including length and robusticity. Ontogenetic data sequences were collected from juvenile and adult archaeological samples of "cold adapted" Inuit and "warm adapted" Khoi San. The results demonstrate statistically significantly different slope between the juvenile and the adult individuals from both samples. Also of interest, the growth trajectories "flex" about the time of adolescence, when linear growth ceases, but somatic growth continues. When Neandertal and early Upper Paleolithic modern humans were contrasted with the extant modern human samples, Neandertals follow a "cold adapted" pattern similar to the Inuit, while the Upper Paleolithic modern humans follow a "warm adapted" pattern.

Given that the Le Moustier specimen had achieved 85% of his adult dental and skeletal growth, this allows us to say something about growth patterns in Neandertals in general. Assuming Le Moustier 1 to be typical of juvenile Neandertals, we can deduce that prior to 85% adult growth, limb proportions, tibial robusticity, and many typical adult Neandertal features were already established. This means that had Le Moustier 1 lived to adulthood, his jaws would have lengthened to increase the extent of mid-facial prognathism, to allow the creation of a retromolar space, and allow the eruption of the third molars as their roots finished forming. As the size and robusticity of the supraorbital torus increased in size, so too would have the size and lateral extent of the frontal sinuses. The size of the Le Moustier 1 femur had yet to reach adult proportions, but presumably this would have occurred by the time of achievement of full adult stature when the long bone epiphyses had completed fusion.

PHYLOGENETIC IMPLICATIONS

Le Moustier 1 is a late Neandertal, dating to approximately 40,000 years or younger (Mellars, 1986; Mellars & Grün, 1991; Valladas et al., 1986) which is about when Upper Paleolithic modern humans appear in Europe. Despite this late date, this individual exhibits features found in specimens from the early part of the Neandertal time range and thus do not suggest any evidence of interbreeding between Neandertals and early modern populations.

Work by JT and AN on aspects of growth and development has shown interesting contrasts in patterns of growth between ancient modern humans and Neandertals:

- Neandertals experienced a delay in postcranial growth relative to dental growth compared to both ancient and modern humans.
- Both Neandertals and Upper Paleolithic modern humans experienced a delay in some cranial measurements relative to dental growth, but only Neandertals exhibit a relative delay in both cranial and postcranial growth.
- Neandertals appear have followed a "cold adapted growth trajectory" while Upper Paleolithic modern humans followed a "warm adapted growth trajectory";
- Both Neandertals and Upper Paleolithic modern humans exhibited an adolescent growth spurt in some skeletal dimensions;

These studies all demonstrate significant contrasts between Neandertals and Upper Paleolithic modern humans suggesting that they followed separate evolutionary pathways.

CONCLUSIONS

The use of the Le Moustier 1 specimen as a case study has allowed us to demonstrate the complementary nature of computerized tomography and traditional morphometric methods. Each technique has unique strengths, allowing us to access particular kinds of data. Furthermore, these methods produce similar results in areas where the techniques overlap (independently confirming each technique). However, the use of CT data and 3-D imagery allows scientists to move into new realms, such as being able to examine the morphology and dimensions of the labyrinth of the inner ear. The other advantages of CT analysis are that it is non-destructive and, using the appropriate software, it allows researchers to view external and internal structures in both 2 and 3- dimensions, to take measurements of those structures, allows the differentiation of real fossilized bone from reconstruction material, and the virtual reconstruction of fragmentary crania. The complementary nature of these techniques is certain to expand our knowledge or basic fossil morphology. These data are critical in the addressing of ontogenetic and phylogenetic research questions.

Acknowledgements

The authors would like to thank Dr Bertrand Mafart and Herve Delingette for the invitation to participate in this special symposium. We would also like to thank Professor W. Menghin, Director of the Museum für Vor- und Frühgeschichte, Berlin for permission to undertake this new reconstruction of the Le Moustier 1 skull and Mrs. A. Hoffmann for her assistance at each stage of this project. Special thanks should go to Professor H. Czichos, the President of the Bundesanstalt für Materialforschung und - prüfung (BAM), for use of their 3-D Microtomograph and software. We are grateful to Dr. J. Goebbels, T. Wolk, and D. Meinel, for their expertise and support throughout the course of this project. Particular thanks should go to Dr. Fred Spoor for his advice and expertise in measuring the labyrinths of the Le Moustier 1 specimen and to Nathan Jeffery for technical assistance. J. L. Thompson is grateful to Dr. Chris Stringer, British Museum of Natural History, for permission to measure comparative cast material and to Mr. R. Kruszynski for his assistance during each of her visits. A. J. Nelson would like to acknowledge the Department of Anthropology, University of Western Ontario.

Addresses of the authors:

J. L. THOMPSON, Department of Anthropology, University of Nevada, Las Vegas, 4505 Maryland Parkway, Box 455012, Las Vegas, Nevada, 89154-5003, USA. *thompsoj@unlv.edu*

A. J. NELSON, Department of Anthropology, University of Western Ontario, London, ON, Canada, N6A 5C2.

B. ILLERHAUS, Bundesanstalt für Materialforschung und - prüfung, Unter den Eichen 87, D-12205 Berlin, Germany.

BIBLIOGRAPHY

BILSBOROUGH, A. & THOMPSON, J. L., In Press a, The Changing Context and Significance of the Le Moustier 1 Neandertal. In *Die Sammlungen von Otto Hauser. Staatliche Museen zu Berlin - Preußischer Kulturbesiz.* edited by A. Hoffmann.

BILSBOROUGH, A. & THOMPSON, J. L., In Press b, Dentition of the Le Moustier 1 Neanderthal. In *The Neandertal Adolescent Le Moustier 1 - New Aspects, New Results,* edited by H. Ullrich. Staatliche Museen zu Berlin - Preußischer Kulturbesiz.

CONROY, G. C. & VANNIER, M. W., 1984, Noninvasive three dimensional computer imaging of matrix filled fossil skulls by high resolution computed tomography. *Science* 226, p. 457-458.

CONROY, G. C. & VANNIER, M. W., 1986, Three-dimensional computer imaging: some anthropological applications. In *Primate Evolution,* edited by J. G. Else & P. C. Lee. Cambridge: Cambridge University Press, p. 211-222.

CONROY, G. C., VANNIER, M. W., & TOBIAS, P. V., 1990, Endocranial features of Australopithecus africanus revealed by 2- and 3-D computed tomography. *Science* 247, p. 838-841.

HAUSER, O., 1909, Découverte d'un squelette du type du Neandertal sous l'abri infériur du Moustier. *L'Homme Préhistorique* 7, p. 1-9.

HEIM, J-L., 1978, Contribution du massif facial à la morphogénèse du crâne Néanderthalien. In Les Origines Humaines et les Époques de l'Intelligence, Paris: Masson et Cie, p. 183-215.

HERMANN, B., 1977, Über die reste des poskranialen skelettes des Neandertalers con Le Moustier. *Z. Morph. Anthrop.* 68, p. 129-149.

HILLSON, S., 1996, *Dental Anthropology.* Cambridge, Cambridge University Press.

HOFFMANN, A., 1997, Zur Geschichte des Fundes von Le Moustier. *Acta Praehistorica et Archaeologica* 29, p. 7-16.

HOFFMANN, A., GOEBBELS, J. & ILLERHAUS, B., 1994, Virtual reconstruction of the skull of Le Moustier - project proposal. Berichtsband 45, Teil 1. *Deutsche Gesellschaft für Zerstörungsfreie Prüfung e.V. 4th Int. Conf. NDT of works of Art.*

HUBLIN, J-J., SPOOR, F., BRAUN, M., ZONNEVELD, F., & CONDEMI, S., 1996, A late Neanderthal associated with Upper Palaeolithic artefacts. *Nature* 381, p. 224-226.

ILLERHAUS, B. & THOMPSON J. L., 1999, Calculating CT data from matched geometries. *DGZfP-Proceedings BB* 67-CD, p. 189-191, (www.dgzfp.de/ctip.html).

ILLERHAUS, B., GOEBBELS, J., & RIESEMEIER, H., 1997, Computerized tomography and synergism between technique and art. In *Selected Contributions to the International Conference on New Technologies in the Humanities and Fourth International Conference on Optics Within Life Science OWLS IV, Münster, Germany, 9-13 July 1996,* edited by D. Dirksen & G. von Bally. Heidelberg: Springer Verlag, p. 91-104.

ILLERHAUS, B., GOEBBELS, J., RIESEMEIER, H., & STAIGER, H., 1997, Correction techniques for detector systems in 3D-CT. *Proceedings of SPIE* 3152, p. 101-106.

ILLERHAUS, B., GOEBBELS, J., REIMERS, P., & RIESEMEIER, H. ,1994, The principle of computerized tomography and its application in the reconstruction of hidden surfaces in objects of art. *4th Inter. Conf. NDT of works of Art DGZFP, Berichtsband* 45, p. 41-49.

KALVIN, A. D., DEAN, D., HUBLIN, J-J., & BRAUN, M., 1992, Visualization in anthropology: Reconstruction of human fossils from multiple pieces. In *Proceedings of IEEE Visualization ,92,* edited by A. E. Kaufman & G. M. Nielson, p. 404-410.

KLAATSCH, H., 1909a, Der primitive mensch der verganenheit und der gegenwart. In *Verhandlungen der Gesellschaft Deutscher Naturforscher und Ärzte,* edited by A. Wangerin. Leipzig: Verlag von F.C.W. Vogel, p. 95-108.

KLAATSCH, H., 1909b, Die neueste ergebnisse der paläontologie der menschen und ihre bedentung für das abstammungs-problem. *Z. Ethnol.* 41, p. 537-584.

KLAATSCH, H., 1909c, Die fortschritte der lehre von der Neandertalrasse (1903-1908*). Ergebn. Anat. Entw.-gesch.* 17, p. 431-462.

KLAATSCH, H., 1909d, Preuves que *l'Homo mousteriensis hauseri* appartient au type du Néandertal. *L'Homme Préhistorique* 7, p. 10-16.

KLAATSCH, H. & HAUSER, O., 1909, Homo mousteriensis hauseri. *Arch. fr. Anthrop.* 35, p. 287-289.

MELLARS, P., 1986, A new chronology for the French Mousterian period. *Nature* 22, p. 410-411.

MELLARS, P. & GRÜN, R., 1991, Comparison of the electron spin resonance and thermoluminescence dating methods: Results of ESR Dating at Le Moustier (France). *Cambridge Archaeological Journal* 1, p. 269-276.

NELSON, A. J. & THOMPSON, J. L., 1999, Growth and Development in Neandertals and other Fossil Hominids: implications for hominid phylogeny and the evolution of hominid ontogeny. In *Growth in the Past: Studies from Bones and Teeth,* edited by R. D. Hoppa & C. M. FitzGerald. Cambridge Studies in Biological Anthropology. Cambridge: Cambridge University Press, p. 88-110.

NELSON, A. J. & THOMPSON, J. L., 2002, Neanderthal Adolescent Postcranial Growth. In *Human Evolution Through Developmental Change,* edited by N. Minugh-Purvis & K. McNamara. Baltimore: John Hopkins University Press, p. 442-463.

NELSON, A. J. & THOMPSON, J. L., In Press, Le Moustier 1and the interpretation of stages in Neandertal growth and development. In *The Neandertal Adolescent Le Moustier 1 - New Aspects, New Results,* edited by H. Ullrich. Staatliche Museen zu Berlin - Preußischer Kulturbesiz.

PONCE DE LEÓN, M. S., ZOLLIKOFER, C. P. E., MARTIN, R. D., & STRINGER, C. B., 2000, Investigation of Neanderthal morphology with computer-assisted methods. In *Neanderthals on the Edge*, edited by C. B. Stringer, R. N. E. Barton, & J. C. Finlayson, Oxford: Oxbow Press, p. 237-248.

SEIDLER, H., FALK, D., STRINGER, C., WILFING, H., MÜLLER, G. B. ZUR NEDDEN, D, WEBER, G. W., REICHEIS, W. & ARSUAGA, J-L., 1997, A comparative study of stereolithographically modelled skulls of Petralona and Broken Hill: implications for future studies of middle Pleistocene hominid evolution. *Journal of Human Evolution* 33, p. 691-703.

SPOOR, C. F. & ZONNEVELD, F.,1994, The bony labyrinth in Homo erectus; a preliminary report. *Courier Forschungs-Institut Senckenberg* 171, p. 251-256.

SPOOR, C. F. & ZONNEVELD, F., 1995, Morphometry of the primate bony labyrinth: a new method based on high-resolution computed tomography. *Journal of Anatomy* 186, p. 271-286.

SPOOR, C. F., WOOD, B., & ZONNEVELD, F., 1994, Implications of early hominid labyrinthine morphology for evolution of human bipedal locomotion. *Nature* 369, p. 645-648.

SPOOR, C. F., ZONNEVELD, F. W., & MACHO, G. A., 1993, Linear measurements of cortical bone and dental enamel by computed tomography: applications and problems. *American Journal of Physical Anthropology* 91, p. 469-484.

TANNER, J. M., 1962, *Growth at Adolescence*. 2nd edition. Oxford: Blackwell Scientific Publications.

TATE, J. R. & CANN, C. E., 1982, High-resolution computed tomography for the comparative study of fossil and extant bone. *American Journal of Physical Anthropology* 58, p. 67-73.

THOMPSON, J. L., In Press, Le Moustier 1 and its place among the Neandertals. In *The Neandertal Adolescent Le Moustier 1 - New Aspects, New Results*, edited by H. Ullrich. Staatliche Museen zu Berlin - Preußischer Kulturbesiz.

THOMPSON, J. L. & BILSBOROUGH, A., 1997, The current state of the Le Moustier 1 skull. *Acta Praehistorica et Archaeologica* 29, p. 17-38.

THOMPSON, J. L. & BILSBOROUGH, A., 1998a, Time for one of the last Neanderthals. In *Proceedings of the XIII Congress of the U.I.S.P.P.- Forlì (Italia, 8/14 September, 1996), volume 2*, edited by F. Facchini, A. Palma di Cesnola, M. Piperno, & C. Peretto. Forlì: Abaco, p .289-298.

THOMPSON, J. L. & BILSBOROUGH, A., 1998b, Time for one of the last Neanderthals. *Mediterranean Prehistory Online* http://www.med.abaco-mac.it/.

THOMPSON, J. L. & BILSBOROUGH, A., In Press, The skull of Le Moustier 1. In *The Neandertal Adolescent Le Moustier 1 - New Aspects, New Results*, edited by H. Ullrich. Staatliche Museen zu Berlin - Preußischer Kulturbesiz.

THOMPSON, J. L. & ILLERHAUS, B., 1998, A new reconstruction of the Le Moustier 1 skull and investigation of internal structures using 3-D-mCT data. *Journal of Human Evolution* 35, p. 647-665.

THOMPSON, J. L. & ILLERHAUS, B., 2000, CT Reconstruction and analysis of the Le Moustier 1 Neanderthal. In Neanderthals on the Edge, edited by C. B. Stringer, R. N. E. Barton, & J. C. Finlayson, Oxford: Oxbow Press, p. 249-255.

THOMPSON, J. L. & ILLERHAUS, B., In Press, 3-D-μCT Endocast and cranial capacity of the Le Moustier 1 skull. In *The Neandertal Adolescent Le Moustier 1 - New Aspects, New Results*, edited by H. Ullrich. Staatliche Museen zu Berlin - Preußischer Kulturbesiz.

THOMPSON, J. L. & NELSON, A. J., 2000, The Place of Neandertals in the Evolution of Hominid Patterns of Growth and Development. *Journal of Human Evolution* 38, p. 475-495.

THOMPSON, J. L. & NELSON, A. J., 2001, Relative postcranial and cranial growth in Neandertals and modern humans. *American Journal of Physical Anthropology*. Special Supplement, p. 149.

THOMPSON, J. L. & NELSON, A. J., In Press a, Estimated age at death and sex of Le Moustier 1. In *The Neandertal Adolescent Le Moustier 1 - New Aspects, New Results*, edited by H. Ullrich. Staatliche Museen zu Berlin - Preußischer Kulturbesiz.

THOMPSON, J. L. & NELSON, A. J., In Press b, The Postcranial skeleton of Le Moustier 1. In *The Neandertal Adolescent Le Moustier 1 - New Aspects, New Results*, edited by H. Ullrich. Staatliche Museen zu Berlin - Preußischer Kulturbesiz.

THOMPSON, J. L., NELSON, A. & B. ILLERHAUS, In press, A study of the Le Moustier 1 Neandertal – summary of results. In *Die Sammlungen von Otto Hauser. Staatliche Museen zu Berlin - Preußischer Kulturbesiz,* edited by A. Hoffmann.

TILLIER, A.-M., 1974, Contribution à l'étude des hommes fossiles Moustériens du Moyen Orient: La pheumatisation de la face. *Paléorient* 2, p. 63-468.

TRINKAUS, E., 1993, *The Shanidar Neandertals*. New York: Academic Press.

VALLADAS, H., GENESTE, J. M., JORON, J. L., & CHADELLE, J. P., 1986, Thermoluminescence dating of Le Moustier (Dordogne, France) *Nature* 322, p. 452-454.

VLCEK, E., 1967, Die Sinus frontales bei europäischen Neandertalern. *Anthrop. Anz.* 30, p. 166-189.

VLCEK, E., 1969, *Neandertaler der Tschechoslowakei*. Prag: Verlag der Tschechoslowakischen Akademie der Wissenschaften.

WEINERT, H., 1925, *Der Schädel des eiszeitlichen Menschen von Le Moustier in neuer Zusammensetzung*. Berlin: Springer.

WIND, J., 1984, Computerized X-ray tomography of fossil hominid skulls. *American Journal of Physical Anthropology* 63, p. 265.

ZOLLIKOFER, C. P. E., PONCE DE LEÓN, M. S., & MARTIN, R. D., 1998, Computer-assisted paleoanthropology. *Evolutionary Anthropology* 6, p. 41-54.

ZOLLIKOFER, C. P. E., PONCE DE LEÓN, M. S., MARTIN, R. D., & STUCKI, P., 1995, Neanderthal computer skulls. *Nature* 375, p. 283-285.

ZONNEVELD, F. W. & WIND, J., 1985, High-resolution computed tomography of fossil hominid skulls: a new method and some results. In *Hominid Evolution: Past Present and Future*, edited by P. V. Tobias. New York: Alan R. Liss, p. 427-436.

ZONNEVELD, F. W., SPOOR, C. F., & WIND, J., 1989, The use of CT in the study of the internal morphology of hominid fossils. *Medicamundi* 34, p. 117-128.

VIRTUAL ANTHROPOLOGY - THE HOPE FOR MORE TRANSPARENCY IN PALEOANTHROPOLOGY

Gerhard W. WEBER

Résumé: Au cours des dernières décennies, une grande quantité de fossiles d'hominidés a été mise au jour. Certains d'entre eux n'ont pas été encore étudiés et d'autres nécessitent d'être ré-étudiés avec des méthodes plus élaborées. Alors que le décompte des restes fossiles est en constante augmentation, il manquait jusqu'à présent une méthode d'inventaire permettant l'acquisition et l'archivage optimale des données de ces spécimens, dont une part est inaccessible à l'observateur comme les structures anatomiques internes. L'Anthropologie Virtuelle (VA) permet des explorations morphologiques tridimensionnelles grâce au traitement informatique des données numériques obtenues à partir des fossiles originaux.

Différentes méthodes comme les scanners médicaux, l'imagerie par résonance magnétique ou les lasers surfaciques, permettent l'acquisition des données tridimensionnelles qui peuvent être secondairement analysées avec des moyens informatiques. Ces données permettent aussi la réalisation de reproduction des spécimens par stéréolithographie. La recherche anthropologique tire un grand bénéfice de ces méthodes qui permettent la visualisation des structures internes. De plus, l'obtention de copies est un moyen de protéger les originaux de toute dégradation. Les analyses morphologiques, à partir de points remarquables ou de surfaces, produit de remarquables résultats grâce aux méthodes de morphométrie géométrique.

Ainsi, l'Anthropologie Virtuelle transforme la Paléoanthropologie en offrant la possibilité d'augmenter notre connaissance de la diversité morphologique des Hominidés, élément clef pour l'analyse des spécimens et l'élaboration des théories phylogéniques. Avec moins de restriction d'accès aux fossiles, une nouvelle étape devient possible : la création d'une base de données tri-dimensionnelles des fossiles récents et des hominidés fossiles qui soit à la disposition de tous.

Abstract: Within the last decades, a remarkable amount of fossil material was excavated, some of it still awaiting a detailed first analysis, some of it requiring re-examination by more developed methods. While the fossil record grew continuously, a methodological inventory evolved to extract critical information about fossilized specimens, most of it preserved in the largely inaccessible interior as unrevealed anatomical structures. Virtual Anthropology (VA) is designed to allow investigations of three-dimensional morphologic structures by means of digital data-sets of fossil and modern hominoids within a computational environment.

3D-data is acquired by different computer-necessitating processes, like CT, MRI, or surface laser scanning. This kind of data also permits the production of accurate 3D-hardcopies of specimens by stereolithographic modeling. Anthropological research profits substantially from methods enabling views to the interior of structures. Additionally, electronic preparation and re-assembly of fossils protects the originals from damage, and quantitative analysis of morphology based on 3D-coordinates of landmarks or surfaces produces biologically meaningful results using Geometric Morphometrics.

As a consequence, Virtual Anthropology is changing the daily routine of anthropological sciences, offering the possibility to extend our knowledge about diversity of hominoids, an imperative component for the assessment of specimens and construction of phylogenetic trees. For more transparency in paleoanthropology, there is the chance for a further step: The foundation of a digital 3D-data archive of recent and fossil hominoids and its opening for global use.

A lot of profound and interesting papers were published within the last years concerning the three-dimensional analysis of paleontological data. In this short contribution, not a specific scientific analysis shall be our focus. Rather I would like to bring a topic of science politics to your attention and introduce a certain perspective of Virtual Anthropology (VA) that can be indeed a hope for paleoanthropology in the future, apart from all the scientific advantages in a narrower sense.

Virtual Anthropology, computer-assisted anthropology, or however else one may call it, is designed to allow investigations of three-dimensional morphologic structures by means of digital data-sets of fossil and modern hominoids within a computational environment. In general, anthropology touches one of the four primary philosophic questions: What is the human being? In asking so, the study of human evolution represents one of the frameworks of our discipline, and thus a potential field of application for VA.

Within the last decades, a remarkable amount of fossil material was excavated, some of it still awaiting a detailed first analysis, some of it requiring re-examination by more developed methods. Morphology and DNA-studies are the main sources of information to distinguish between paleospecies, but variation is often poorly known. Partly, because there are so few specimens, but partly because it can be also very difficult to gain access to the existing fossils record.

What are the main goals of Virtual Anthropology?

1. Acquire new qualitative traits, especially so far inaccessible ones (e.g., endocranial traits, sinuses, medullary cavities, etc.).

2. Measure and compare quantitative traits (exo & endo) with high accuracy and reproducibility, and to study variation

3. Provide digital 3D-data of recent and fossil hominoid material and the relevant methods of analysis.

A main issue for paleoanthropologists is the definition of a paleospecies. Fundamental is a profound knowledge of variation and the inclusion of a multitude of traits. Of course, also those structures that are hidden should be studied. When a scientist has a new idea for a comparative analysis, it can be really troublesome to gather all the material of the comprised specimens that are physically stored around our planet. If this idea involves traits of hidden structures too, the only way to investigate them is to use Virtual Anthropology anyway. Perhaps, once you were tempted to think about checking a measurement done by another author in his recent paper? You'll have fun flying to Johannesburg, Nairobi, or Beijing.

Once a specimen is CT-scanned, the resulting virtual objects have a great advantage: They are resting on the disk of your personal computer, waiting there 24h a day, 365 days a year. No more applications for access are necessary. You just have to login and start your software. Virtual objects can be rotated, scaled, moved, cut or imaged. The comparison of a real object with its virtual copy shows that there is only a negligible difference (Hildebolt et al., 1990; Richtsmeier et al., 1995; Feng et al., 1996; Weber et al., 1998), except a lack of texture information.

There is also the possibility to bring back virtual objects to reality. Such "stereolithographic models" are of significant importance for the investigation of morphology (Seidler et al., 1997). But sometimes, their relevance is misunderstood. They are not meant as a substitute for traditional casts, they are an addition. They provide some advantages over traditional casts like the possibility to produce a model in parts and to look at the structure of cavities, a task that is not grantable with usual casts. Moreover, there is the possibility to produce parts of anatomical structures on an enlarged scale, let us say the orbit or the maxilla on a scale of 1:10, and to produce electronically casted cavities also as physical objects, like the labyrinth or the frontal sinus.

Virtual Anthropology is THE gateway to hidden anatomical structures and allows to penetrate into the last corner of a specimen. In Figure 1 are just a few examples, like an endocast of a frontal sinus, including volume computation, the study of the tooth roots of an *A. robustus* specimen, or an example of computing thicknesses along the surface of a single bone and graph them as topographical thickness maps (Weber et al., 2000).

For the comparison of the shape of organisms it is necessary to utilize quantitative methods. The traditional morphometry mainly uses distance and area measurements, whereby the original shape is not recoverable from the usual matrices of measurements, even if multivariate statistical methods are used. Geometric morphometric methods are using coordinates, and are permitting to distinguish between shape- and size-variation as well as to analyze which structures were changing or remaining constant (Bookstein et al., 1999). Traditional orientation problems (e.g., Frankfurt plane) are eliminated by superimposition methods, for example procrustes. Finally, thin plate spline (Figure 2) can visualize the change of the position of landmarks as deformation grid. Each analyses can start from the original CT-scans because landmark coordinates are extractable from this kind of data.

Fossils undergo a series of processes along their way from biosphere to lithosphere and can be embedded into sediments or have calcite incrustations. In each case, the anatomical investigations are affected and a preceding preparation is unavoidable. Virtual Anthropology offers the possibility of electronic preparation which is reversible and avoids physical contact with the original specimen. In some cases, the incrustations have a different gray value than the bone and their elimination is quiet easy by setting the appropriate threshold. In other cases, there is an overlap in the density profiles of the incrustations and the bone. In such a case, more sophisticated methods, involving various filter algorithms and manual processing of each CT-slice (Prossinger et al, 1998), are needed but can be worth the effort if one obtains an electronically cleaned specimen that allows subsequent analyses of its endocranial morphology.

One of the most challenging tasks for VA is the reconstruction of fragmented fossils and the reversal of deformations. Missing features can be re-created by mirroring pieces or by completing with pieces from similar specimens. The production of such a composite fossil does not affect the precious original, and the electronic manipulations are easier to reproduce than physical ones. But most of the reassembling experiments still rely more on the "morphological eye" of the scientists than on empirical standards. Statistical information on the distribution of homologous landmarks, ridge curves, and other surface properties is indispensable. We are currently working on the development of parameterized skull models (Weber & Neumaier, 2001) which will allow to reconstruct parts to a certain degree of reliability.

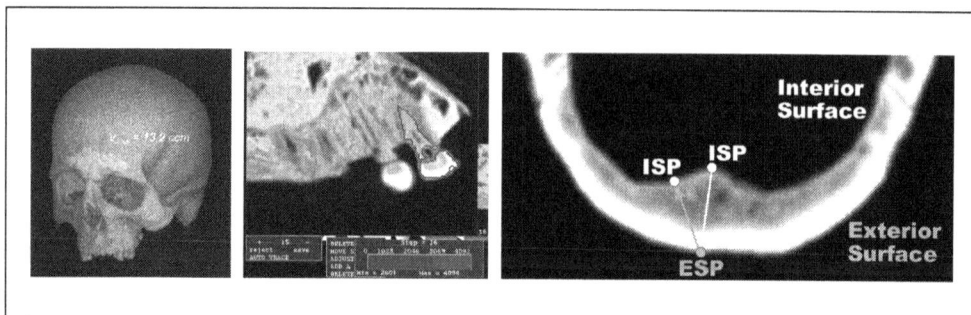

Figure 1 - Visualization and computation of volumes (left), study of tooth roots (middle), semiautomatic thickness measurements (right)

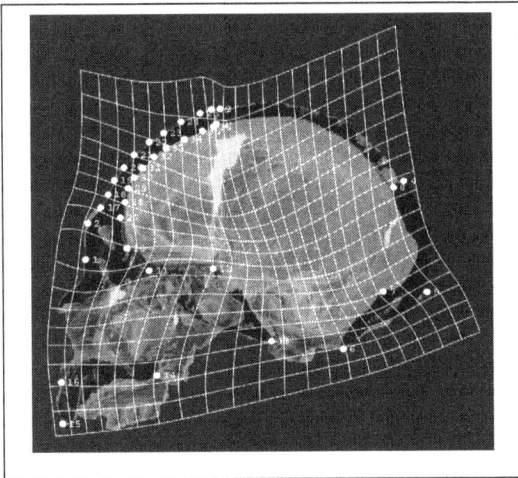

Figure 2 - Thin-plate spline deformation grid
(*Homo sapiens* ⇒ *Homo heidelbergensis*)

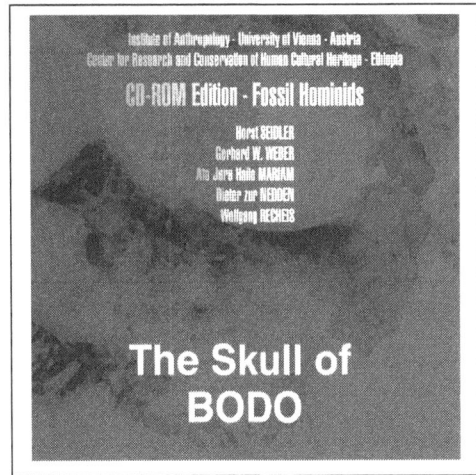

Figure 4 - The first CD-ROM with 3D-data of a fossil hominid

In Figure 3 an example for a simple, non-linear model based on conic triangles is shown. We use traditional measurements, Cartesian coordinates and differential-geometric data, whereby the latter and semi-landmarks can only be obtained on the virtual object and not on the physical one.

Undoubtedly we made a big step forward in paleoanthropology within the last decade. But something is still unsatisfactory. **WHAT IS IT?**

The number of fossils is growing rapidly, the potential to analyze the material is growing too. But the access to the fossils is still very limited. I think it is time to implement a democratic process for paleoanthropology. I call this idea a "Glasnost" in paleoanthropology (Weber, 2001), a Glasnost, if all the discovered fossils would be easily accessible. It is increasingly obvious that knowledge about diversity is the key for the assessment of specimens and the constructing of phylogenetic trees. But how can morphologic diversity be studied on a large scale if access to fossil specimens is restricted by time, by distance, or simply by the benevolence of curators? Of course, the latter have the responsibility to protect their treasures and are alarmed by the increasing

number of applications for access. In some cases, conventional distance measurements are taken for the umpteenth time, always risking micro-destructions due to contact with sharp instruments. That can be easily avoided using digital data.

Hominoid fossils are the heritage of all mankind. The digital 3D-data of all fossils should be freely accessible for global use. Electronic archives, on the Internet or on CD-ROM, are accessible at any time from the desktop. We made a beginning in this direction by publishing the first CD-ROM with 3D-data of a fossil (Figure 4), the Bodo cranium, in 1999 (Seidler et al.), and we are planning to publish more specimens. For the internal use at the institute, currently we run a trial version of an electronic archive of modern hominoids, and clearly, we are not alone. Others also have such parts of a global archive on their servers. Much work needs to be done to organize and administrate such an archive, for example, the quality standard, the distribution mode, or the financial aspect have to be worked out. But I hope that more people will realize the advantages for our field of science, including the chance to present a new picture of our science to the public. My suggestions for discussion are:

- We should start with the foundation of a global electronic archive for skeletal morphology of fossil and recent hominoids, comparable to the Human Genome Project.

- All fossils should be digitized with standardized methods. If an author wants to publish the description of a new fossil she or he should be obliged to publish the 3D-data too. If a fossil is announced but not officially published within a certain time limit, it should be scanned and the data also be published in the archive.

- The electronic archive should be opened.

High resolution CT-scans using m-CT enlarge the spectrum of potential investigations (Thompson & Illerhaus, 1998) and could be the future standard. Of course, the computational costs are enormous and at the moment, only a few institutes can afford to work with this technology. But with the steady increase in computer power this should not be an insoluble problem for a long time. With such a volume data set, a

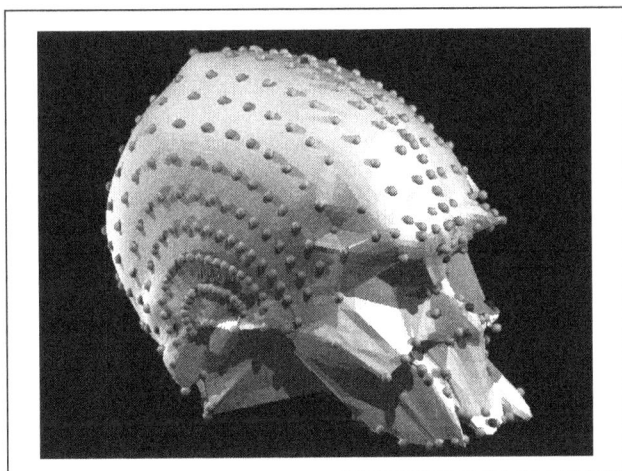

Figure 3 - Example of a simple parametric skull model
(conic triangles)

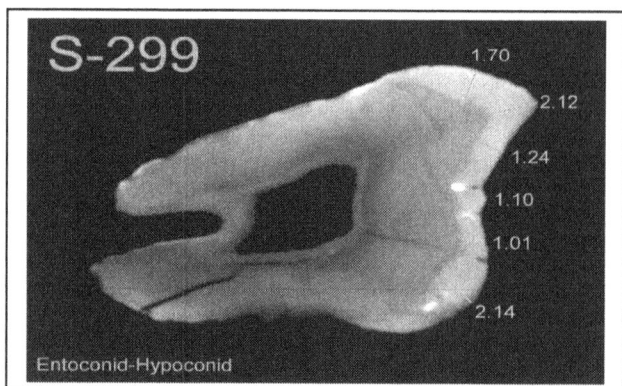

Figure 5 - Measurement of the enamel thickness; resolution is 60 μm (with the help of Dr. Bernhard Illerhaus, Bundesanstalt fuer Materialforschung und –pruefung, BAM-Berlin)

BIBLIOGRAPHY

BOOKSTEIN FL, SCHAEFER K, PROSSINGER H, SEIDLER H, FIEDER M, STRINGER C, WEBER GW, ARSUAGA JL, SLICE DE, ROHLF F, RECHEIS W, MARIAM J, MARCUS L, 1999, Comparing Frontal Cranial Profiles in Archaic and Modern Homo by Morphometric Analysis. *Anat. Rec.* 257:217-224.

FENG Z, ZIV I, RHO J, 1996, The Accuracy of Computed Tomography-Based Linear Measurements of Human Femora and Titanium Stem. *Investigative Radiology* 31/6:333-337.

HILDEBOLT CF, VANNIER MW, KNAPP RH, 1990, Validation Study of Skull Three-Dimensional Computerized Tomography Measurements. *Am. J. Phys. Anthropol.* 82:283-294.

PROSSINGER H, WEBER GW, SEIDLER H, RECHEIS W, ZIEGLER R, ZUR NEDDEN D, 1998, Electronically aided preparation of fossilized skulls: Medical imaging techniques and algorithms as an innovative tool in palaeoanthropological research. *Am. J. Phys. Anthropol. Suppl.* 26:181.

RICHTSMEIER JT, PAIK CH, ELFERT PC, COLE TM, DAHLMAN HR, 1995, Precision, Repeatability, and Validation of the Localization of Cranial Landmarks Using Computed Tomography Scans. *Cleft Palate-Craniofacial Journal* 32/3:217-227.

SEIDLER H, FALK D, STRINGER C, WILFING H, MUELLER G, ZUR NEDDEN D, WEBER G, RECHEIS W, ARSUAGA JL, 1997, A comparative study of stereolithographically modeled skulls of Petralona and Broken Hill: implications for future studies of middle Pleistocene hominid evolution. *J. Hum. Evol.* 33:691-703.

SEIDLER H, WEBER GW, MARIAM AJH, ZUR NEDDEN D, RECHEIS W (eds.), 1999, *The skull of Bodo. CD-ROM Edition - Fossil Hominids*, Vienna: Inst. of Anthropology, University of Vienna, http://www.anthropology.at/bodo/bodo.html.

THOMPSON JL, ILLERHAUS B, 1998, A new reconstruction of the Le Moustier I skull and investigation of internal structures using 3-D-uCT data. *Journal of Human Evolution* 35:647-665.

WEBER GW, RECHEIS W, SCHOLZE T, SEIDLER H, 1998, Virtual Anthropology (VA): Methodological Aspects of Linear and Volume Measurements - First Results. *Coll. Antropol.* 22:575-583.

WEBER GW, KIM J, NEUMAIER A, MAGORI CC, SAANANE CB, RECHEIS W, SEIDLER H, 2000, Thickness Mapping of the Occipital Bone on CT-data - a New Approach Applied on OH 9. *Acta Anthropologica Sinica Suppl. Vol.* 19:37-46.

WEBER GW, NEUMAIER A, 2001, *Morphological comparison and fragment re-assembly of fossil and modern skulls using parameterized reference models*. Austrian Science Foundation, Project no. P14738, Vienna.

WEBER GW, 2001, Virtual Anthropology (VA): A Call for Glasnost in Paleoanthropology. *Anat. Rec. (New Anat.)* 265/4: 193-201.

scientist can decide, after having explored the 3D surface properties of the cranial vault, for example to measure the volume of the hypothalamic pit and determine the radii of the semicircular canals, or to study dental enamel thickness or even perikymata structures of the teeth. In Figure 5 you see a m-CT of a third molar of an *A. afarensis*, done with the help of the BAM in Berlin on their self-constructed scanner. The exact measurement of enamel thickness is easy to perform, which leads at the end to hard numbers and not to descriptions like "thin-enameled" or "thick-enameled" which you might find in most of the literature.

Finally, another aspect: There is a gigantic NIH-project with the goal to provide anatomical data for all universities in the United States ONLINE. Our institute is the first test site outside the USA, being invited to contribute the development. The data is based on the Visible Human Project, a 47 GB data volume of kryosections. The coming user of the developed software will be able to explore the complete body in three dimensions online via Internet. Support will be given by descriptions of anatomical structures that appear after pointing at them. All kinds of measurements will be possible. Within the next two years, the development should be completed.

Meanwhile, we anthropologists should also think about the opportunity to use such a tool for a new way of teaching the morphology of fossil specimens because the software is easily adaptable for this kind of application. Students would have online access to fossils to train evolutionary anatomy – certainly also a contribution to more transparency in paleoanthropology.

Author's address:

Gerhard W. Weber
Institute for Anthropology
University of Vienna
Althanstr. 14, A-1090 Vienna
Austria
http://www.anthropology.at

COMPARISON OF A THREE-DIMENSIONAL AND A COMPUTERIZED-ASSISTED METHOD FOR CRANIO-FACIAL RECONSTRUCTION: APPLICATION TO TAUTAVEL MAN

Guillaume ODIN, Gérald QUATREHOMME, Gérard SUBSOL, Hervé DELINGETTE,
Bertrand MAFART, Marie-Antoinette de LUMLEY

Résumé: Les auteurs ont tenté une reconstitution crânio-faciale de l'Homme de Tautavel, daté d'environ 450. 000 ans, découvert par H. de Lumley en 1971, à partir d'un moulage de la face et du pariétal de ce fossile, associés à des moulages, pour les os manquants, provenant d'Homo erectus découverts dans d'autres sites. Une reconstitution manuelle a été effectuée avec les méthodes utilisées en médecine légale pour l'identification des corps. Une méthode informatique tri-dimensionnelle a également été utilisée, en collaboration avec les chercheurs de l'INRIA (projet Epidaure, Sophia-Antipolis, France). Les résultats obtenus montrent que les proportions sont globalement respectées par les deux méthodes alors que la ressemblance entre les deux reconstitutions n'est pas parfaite. Les auteurs concluent à l'interêt d'utiliser conjointement ces méthodes pour obtenir une reconstitution des hommes préhistoriques.

Abstract: The authors report on the craniofacial reconstruction of the Tautavel Man 450.000 BP, discovered by H. de Lumley in 1971, based on an entire skull built with the front and the parietal bone of the Tautavel Man and fragments that were found at Homo erectus other sites. A three-dimensional manual craniofacial reconstruction has been set up for identification of missing persons in actual forensic cases. A three-dimensional computerized-assisted method was also settled in collaboration with, and by the INRIA team («Epidaure group», Nice Sophia-Antipolis, France). Both methods have been applied to the cranio-facial reconstruction of the Tautavel Man. The results show that global proportions are respected between both methods, whereas the resemblance between the non computerized- and the computerized-assisted method is not obvious. The conclusion is that both methods may be useful for obtaining a representation of prehistoric men.

Identification of human remains is a huge forensic issue. Identification records are often missing, as dental charts or DNA databases, and the last and ultimate approach may be craniofacial reconstruction, especially in difficult cases as decomposed or skeletonized remains. In turn facial reconstruction has emerged as an increasingly important tool in forensic pathology and forensic anthropology, generating a lead for positive comparative identification. These sophisticated techniques may be 2D or 3D, computerized-assisted or not.

In this work we will describe two techniques that we have settled for forensic practice (e.g. 3D-manual technique, and 3D-computer-assisted reconstruction), and we will indicate the results we have obtained with both techniques on the "Tautavel man" skull.

METHODS

Three Dimensional Manual Facial Reconstruction

The most traditional 3D approach is the manual build-up of the skull with a clay or clay-like material. This manual "plastic", or "sculptural" technique consists of applying a clay-like substance onto the skull, through the relationship between bone and soft tissues (Fedosyutkin et al, 1993). Skull morphology and metric features must be assessed before placing marks on precise anthropological points, where the average tissue depths are known according to the age, sex, and race of various populations. Then the space between these points are filled in with clay. Some areas such as ears, eyes,

nose, mouth and lips are difficult to settle because they do not have clear bony indicators.

Three Dimensional Computerized Methods

The computerized 3D craniofacial reconstruction has been developed by only a few scientists (Evenhouse et al, 1992, Quatrehomme et al, 2000). We have used a method that was set up by a cooperation with the Epidaure group of the Inria team, in Sophia-Antipolis (France). The whole method is available in (Quatrehomme et al, 1997). We have used a first set, consisting of a skull model S1 and a facial model F1, which is the unknown face that we have to reconstruct, whereas the second set (Sr, Fr) is the reference head with a known skull and its known face. The method of reconstruction consists in deforming the surface of the skull, through a global parametric algorithm T, so that :

$$T (Sr) = S1$$

and this transformation is applied to the reference face Fr, with the issue of :

$$T(Fr) = F1?$$

This transformation is based only on some salient lines ("crest lines"), which correspond mathematically to lines of absolute maxima of the largest principal curvature.

RESULTS

3D manual- and 3D computerized methods had been developed for forensic practice. The validation of the manual

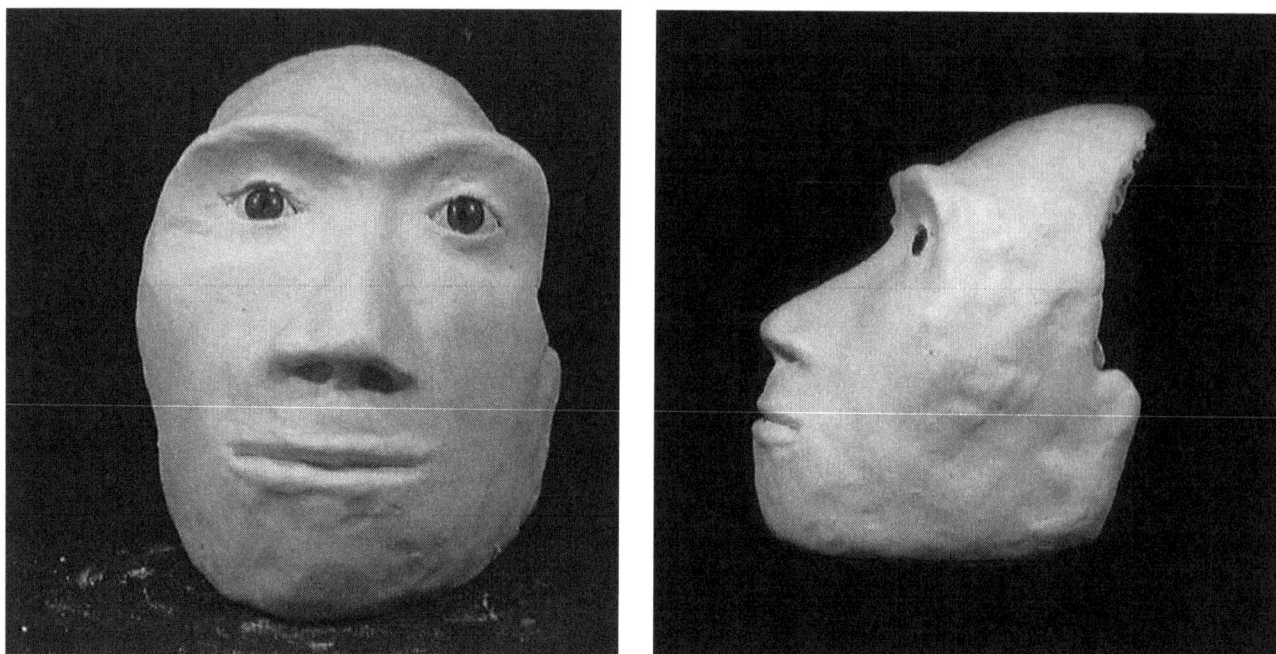

Figure 1 - Final result both methods

method has been achieved. Globally, out of 24 controlled cases, the results in terms of resemblance were considered as good results in 4 cases, middle results in 5 cases, and poor results in 15 cases (Quatrehomme, 2000). From a forensic point of view, it means that there is a hope for the family to recognize the missing person in nearly 40% of the cases. The validation of the computerized-assisted method is currently in process.

In the present work we have applied both methods on the Tautavel man (photography 1, 2), discovered by Henri de Lumley in 1971 and considered as dating back to 450 000 BP, on an entire skull cast built with the face and the parietal bone of the Tautavel Man and fragments that were found at Homo erectus other sites.

DISCUSSION

The 3D manual approach is very popular. Conversely, 3D computerized methods are sparsely described in the literature. Only a few of the computed facial reconstruction programs are true 3D.

But both methods suffer of a lack of scientific validation (Quatrehomme et al, 1994). This evaluation must deal with the relationship between the main anthropological points and the soft tissue depth at these locations, but also the relationship of various areas of the face between each others. But the validation must concern the issue of resemblance (which is the main issue in forensic practice) as well (Quatrehomme, 2000).

The 3D manual craniofacial reconstruction of the Tautavel man cannot be considered as a scientific one, because we do not know the average tissue thicknesses of the main anthropological points of the Tautavel man.

The 3D computerized method do not use this average of soft tissue depth. It is based on another approach which a 3D deformation of skull into another, and the application of this transformation to the reference face into the face to reconstruct (here the Tautavel man). Unfortunately we have no mean to check the results on ancient skulls. Furthermore we have observe that the resemblance between both methods is not obvious. Nevertheless, because it was the conclusion of our previous controlled study (Quatrehomme et al, 1997, Quatrehomme, 2000), we make the hypothesis that these methods preserve the relations between the bone and the face in various areas of the face, and preserve the balance between these areas. Even if the final result does not actually look like this peculiar and unique Tautavel man (in terms of precise resemblance), it is probably close to it, and this information is interesting in paleoanthropology. In conclusion we may assume that both methods are useful for obtaining a representation of prehistoric men.

Authors' addresses:

Guillaume ODIN, Gérald QUATREHOMME,

Laboratoire de Médecine Légale et Anthropologie médico-légale,

Faculté de Médecine, Avenue de Valombrose,

06107 Nice cedex 2, France.

Gérard SUBSOL,

Laboratoire d'Informatique, Université d'Avignon, France

Hervé DELINGETTE,

Projet EPIDAURE, Institut National de Recherche en Informatique et en Automatique, Sophia Antipolis, France

Bertrand MAFART, Marie-Antoinette de LUMLEY

Laboratoire d'Anthropologie, Faculté de Médecine, UMR 6569, Université de la Méditerranée, Marseille, France

Correspondance to: Gérald QUATREHOMME, quatrehomme.g@chu-nice.fr

BIBLIOGRAPHY

EVENHOUSE, R., RASMUSSEN, M., SADLER, L., 1992, Computer-aided forensic facial reconstruction. *J Biocommunication* ,19, 2, 22-28.

FEDOSYUTKIN, B.A., NAINYS, J.V.,1993, The relationship of skull morphology to facial features. In : ISCAN,M.Y., HELMER,R.P., (eds), *Forensic analysis of the skull*, Wiley-Liss, New York, pp. 199-213.

QUATREHOMME, G., GARIDEL, Y., BAILET, P., LIAO, Z.G., GRÉVIN, G., OLLIER, A.,1994, An attempt of scientific evaluation in three-dimensional facial reconstruction : preliminary results. *XVIth Congress of the International Academy of Legal Medicine and Social Medicine*, Strasbourg (France), 31 May-2 June ,

QUATREHOMME, G., COTIN, S., SUBSOL, G., DELINGETTE., H, GARIDEL, Y., GRÉVIN, G., FRIDRICH, M., BAILET, P., OLLIER, A., 1997, A fully three-dimensional method for facial reconstruction based on deformable models. *J Forensic Sci*, 42, 4, 647-650.

QUATREHOMME, G., ISCAN, M.Y.,2000,Forensic facial reconstruction. *Encyclopedia of Forensic Sciences* . JA SIEGEL, SAUKKO,P.J., & GC KNUPFER,G.C.,(eds.). Academic Press

QUATREHOMME , G., 2000, Reconstruction faciale : intérêt anthropologique et médico-légal. Thèse Sciences (Anthropologie), Université de Bordeaux, 6 Juin 2000.

HOMINID TOOTH PATTERN DATABASE (HOTPAD) DERIVED FROM OPTICAL 3D TOPOMETRY

Ottmar KULLMER, Mathias HUCK, Kerstin ENGEL, Friedemann SCHRENK,
Timothy BROMAGE

Résumé: HOTPAD est une plate-forme d'acquisition des données tridimensionnelles dentaires. Un système de topométrie optique portable a été utilisé pour réaliser l'acquisition en haute résolution des surfaces de dents des premiers espèces du genre Homo. Dans cette étude, nous présentons le concept et la structure du système HOTPAD à partir de l'analyse d'une deuxième molaire d'un hominidés de Java, Indonésie, conservée au Forschungsinstitut Senckenberg de Francfort. Des modèles de couronne dentaire reproduites en réalité virtuelle ont été utilisés pour les calculs morphométriques. Nous présentons des méthodes permettant l'extraction et la visualisation des distances, surfaces et volumes quelque soit la forme des couronnes dentaires. De plus, d'autres paramètres comme la forme des relief, la surface des cuspides, les arrondis et les angles sont analysés pour montrer l'étendue des possibilités de ces scanners surfaciques à haute résolution pour l'étude de la morphologie des molaires et de l'usure dentaire.

Abstract: HOTPAD is a platform for providing three-dimensional data of teeth. A portable optical topometry system is used here to acquire high-resolution point clouds of tooth surfaces of early Homo species. In this contribution we introduce the concept and structure of HOTPAD, on the basis of the early hominid second molar assemblage from Java, Indonesia, housed in the Forschungsinstitut Senckenberg in Frankfurt. Enhanced Virtual Reality models of molar crowns are used for morphometric calculations. Here we present methods to extract and visualize distance, area and volume parameters of overall shape of tooth crowns. Furthermore functional parameters such as relief shape, cusp area, sloping and angle are documented to explore possibilities of high-resolution surface scans for studying molar morphology and wear pattern.

CONCEPT OF HOTPAD

The idea of HOTPAD arose during a pilot study on digital documentation of tooth morphology. We tested an optical 3D topometrical system to acquire data of tooth crowns for the purpose of occlusal surface analysis. During this test it became clear that high-resolution virtual models would provide an efficacious means of obtaining precise metric data of teeth at a significant savings of time and effort.

HOTPAD is a quantitative image database of teeth based on high-resolution Virtual Reality (VR) models. It is an html-programmable platform to enable data acquisition from specimens for comparison and presentation. It can be used as a simple catalogue or an information centre, but it is also intended to be a forum for the discussion of new ideas and methods concerning tooth pattern analyses. Interested individuals are invited to view and survey low-resolution vrml-models, using a viewer such as Cosmoplayer or Cortona. Polygon files in higher resolution are also available for downloading. Colleagues can evaluate specimens in the collection for their own research purposes.

The first version of HOTPAD (Figure 1) contains information about teeth of early *Homo* species recovered by G.H.R. von Koenigswald from the island of Java, Indonesia. Material from the Sangiran collection housed in the Forschungsinstitut Senckenberg in Frankfurt am Main consists of three fragmentary skulls (Sangiran 2,3 and 4), one maxilla, six mandible fragments (Sangiran 1, 4, 5, 6a, 6b and 6c) and 52 single teeth (Sangiran 7.x) (von Koenigswald, 1940, 1950, 1954; Grine & Franzen, 1994). Because the stratigraphic

context of most of these finds is not known with sufficient reliability, evolutionary researchers often neglect these specimens. Despite this caveat, in 2001 we began a long-term project on the reconstruction and quantification of early *Homo* tooth morphology with both the intentions of bringing this neglected sample to light and to develop and put into practice our 3D topometric method. The main goal is to recognize and interpret evolutionary trends in tooth shape development and change, such as megadonty in earliest *Homo*.

We describe here the technique of optical 3D topometry and methods of data extraction from virtual models. Measurements and basic comparisons of structural parameters are also available. New structural parameters are introduced to quantify and understand the occlusal surface of teeth and specific changes arising through wear.

OPTICAL 3D TOPOMETRY

For 3D-data acquisition we used the topometrical 3D-measurement system OptoTop (Breuckmann GmbH, www.breuckmann.com; Figure 2). The OptoTop 3D-sensor works on the principal of optical triangulation, based on a simple mathematical equation. Knowing the sensor geometry of the camera CCD makes it possible to calculate for each point at x and y, the height (z). The measuring method is refined through the combination of Gray-Code and Phaseshift techniques (Breuckmann, 1993). In the Gray-Code technique a special projector produces a sequence of fringe patterns of various grating periods on the object (Figure 3) and a CCD-

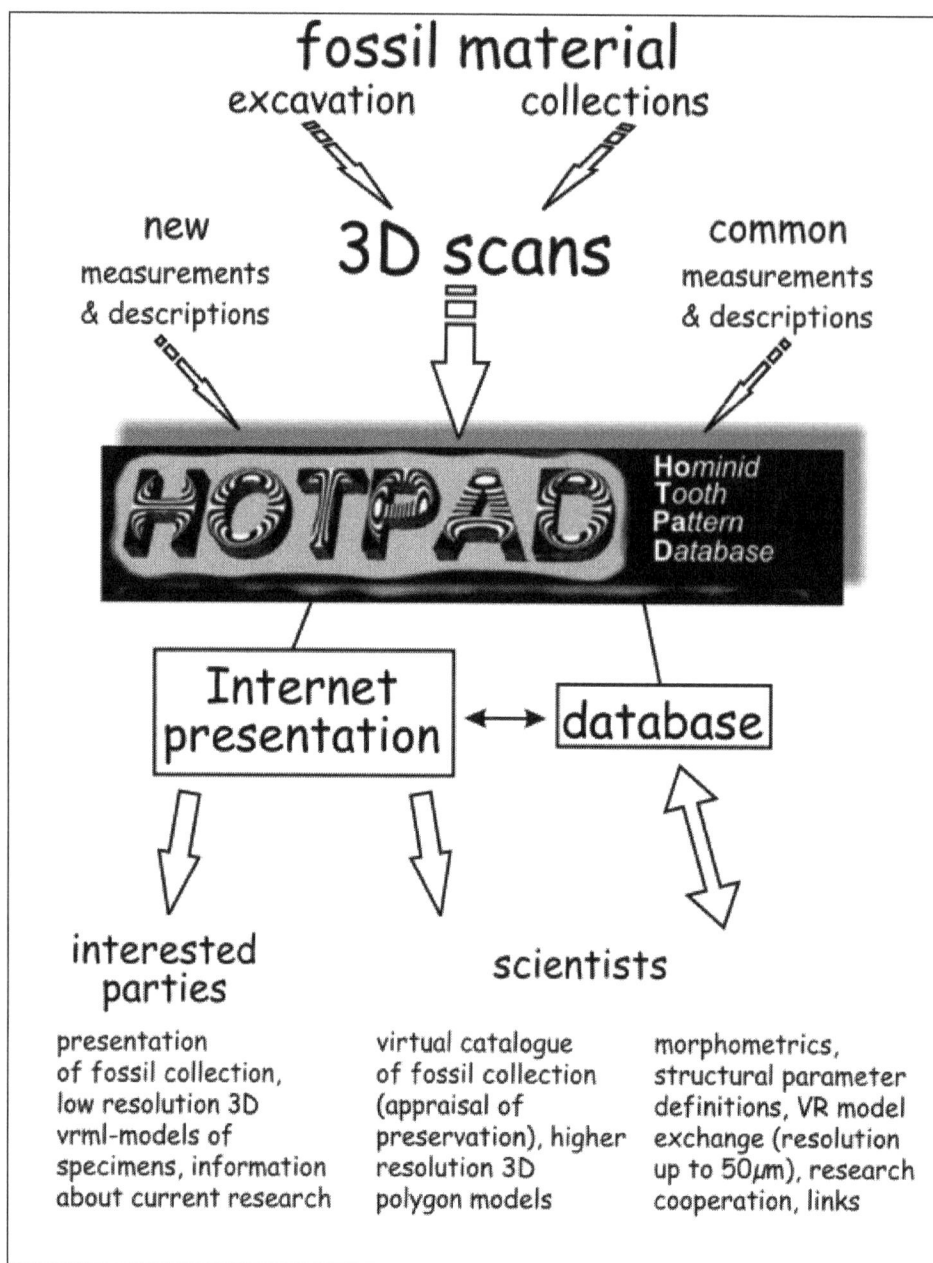

Figure 1 - HOTPAD is a virtual catalogue of fossil hominid teeth, including morphometric data of surface analyses. It contains virtual 3D surface models of specimens from the famous site of Sangiran on the island of Java, Indonesia. Interested parties and scientists all over the world can use HOTPAD as an information centre and source of data accessible via Internet.

camera (1x1.3k chip size), located at a certain angle towards the direction of illumination, records the object (e.g. tooth), simultaneously with each of the various projected fringes. After the Gray-Code pattern images have been acquired, the projector produces very fine sinuous stripe pattern onto the object, which is shifted about 90° after each image capture. A sophisticated computer calculation permits the extraction of x,y,z coordinate of each image point, through superimposition and analysis of each image. The 3D coordinates can be calculated pixel by pixel by means of these object patterns according to rules of triangulation.

In order to have a complete virtual model of a tooth we record the 3D-coordinates from several views. The number of views depends on the relief height, surface reflection, and complexity of the object. Experience shows that fossil tooth recordings require more image perspectives than modern teeth in order to have a reasonably complete model. Large differences in contrast and reflection render fossil surfaces more difficult to image, thus requiring individualized scan-parameter settings (e.g. repetition of the image sequence with several light intensity and contrast threshold settings). Powdering techniques using, for instance, ammonium chloride powder to evenly whiten the surface, can be very useful when variation in image contrast is high. Our experience with single teeth has indicated that between six and ten views are required before having a suitable model. Once all views are acquired they are aligned with PolyWorks, a modular software package from Innovmetric (www.innovmetric.com).

To obtain high-resolution point clouds we scan specimens with a measuring volume of 60x80x50 mm, producing for

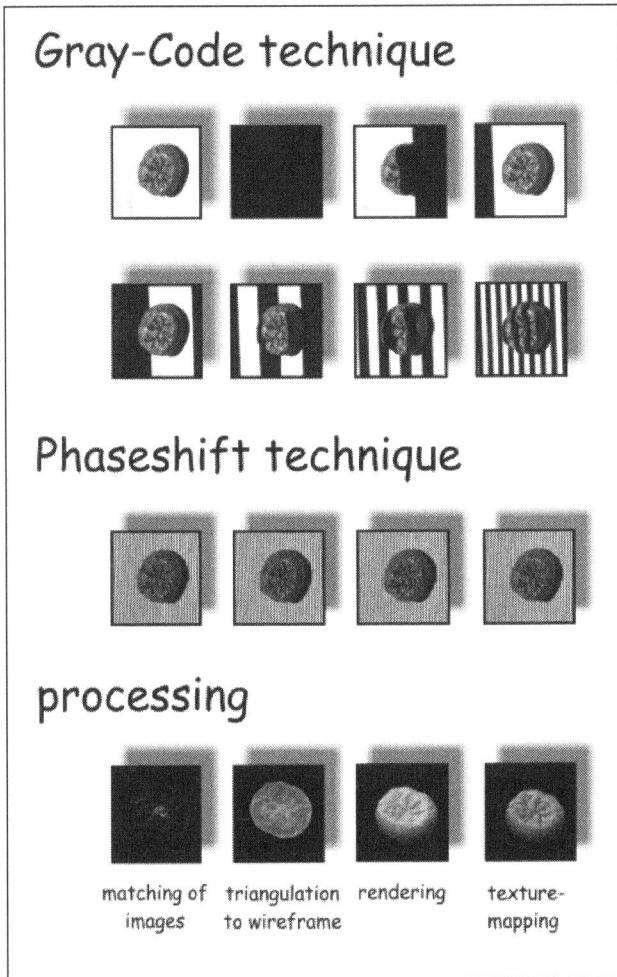

Figure 2 - The portable 3D system OptoTop consists of a CCD Camera and a fringe pattern projector. Calculation of 3D coordinates for each object point, recorded on the CCD chip, is based on the principle of triangulation. The distance d can be calculated, if the exact space between the CCD camera and projector (a), and the angle (α), are known.

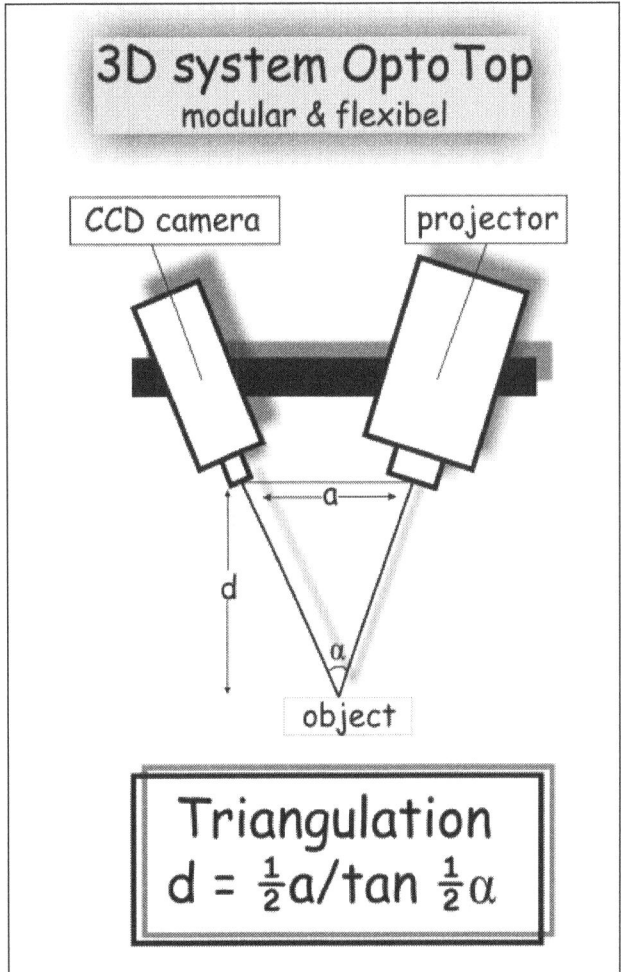

Figure 3 - The projector sends strip pattern with increasing number of lines on the object, while the camera is recording each image (Gray-Code technique). A high resolution point cloud results after the combination of Gray-Code and Phaseshift technique (projection of a sine shaped pattern and four times shifted about 90°) and a rather sophisticated computer calculation of the image stack. Further processing of the scans includes matching of images, triangulation to a wire frame model, rendering and original texture mapping.

each scan a maximum of 1.3 Million points. The resulting point clouds are matched from all views with a basic point grid of 50μm and an accuracy of +/- 25μm. We then import the point clouds into the editing, analysis, and visualisation modules of PolyWorks. The integrated triangulation tool calculates surfaces based upon three neighbouring points of the cloud (polygon meshing) (Figure 3). The surface of a tooth is thus rendered and can be studied with various measurement and analysis tools. Furthermore, because the topometric system is recording grey values for each point, the VR model is rendered with realistic surface contrasts (Figure 4).

Limits of three-dimensional surface rendering occur when the object shows extreme cavities or curvature and no angle position of the sensor can be found to illuminate object boundaries. Nevertheless pilot studies with different objects, like mammal teeth, skulls, long bones, fossil brain casts, as well as fossil and modern footprints (Kullmer et al., in press) have demonstrated that optical 3D-topometry is of multipurpose use in palaeontology and archaeology.

TOPOGRAPHIC RELIEF ANALYSIS OF TEETH

For more than 100 years palaeontologists have analysed teeth, particularly because many extinct species are only known from their dentitions. Teeth are a source of information about species affinity, dietary preferences and function, the environment, and evolutionary trends in tooth pattern development.

The chewing process works with extreme precision. Any deviation of occlusion from normal results in reduced effectiveness. Mastication is influenced by many factors such as jaw movement and the primary topography, or relief, of the teeth (primary relief). The spatial position of a tooth is also a limiting factor, determining how deep a specific cusp may dip into the corresponding basin of the occluding tooth. During the mastication cycle contacts occur between food

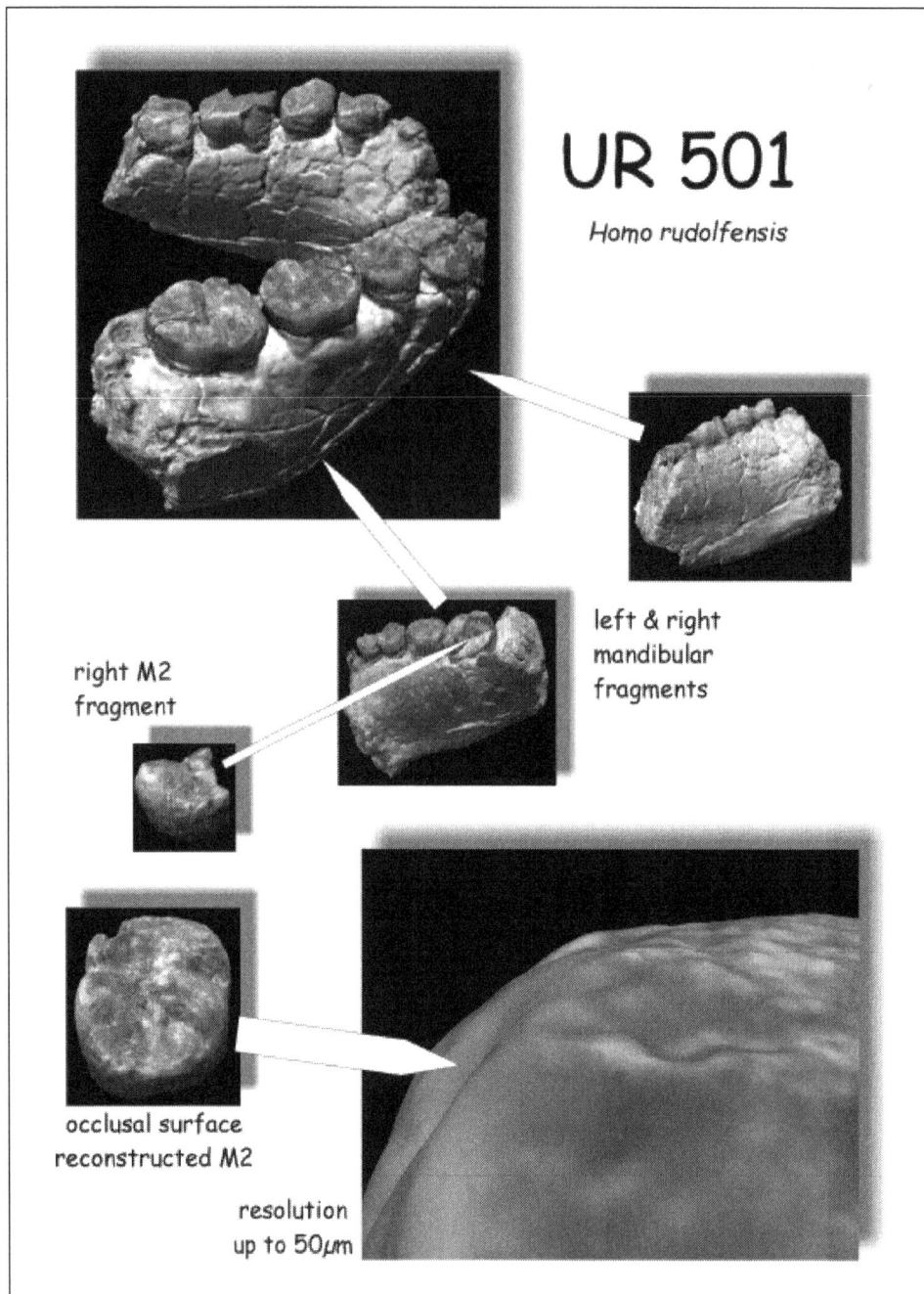

UR 501

Homo rudolfensis

right M2
fragment

left & right
mandibular
fragments

occlusal surface
reconstructed M2

resolution
up to 50μm

Figure 4 - Rendered virtual 3D surface model and reconstruction of mandible fragments
(e.g. early *Homo rudolfensis* jaw UR 501 from the Chiwondo Beds in northern Malawi)
derived from optical 3D topometry.

and enamel, and between upper and lower teeth. From the beginning of a tooth's functional life, chewing efficiency increases as its crushing, grinding and shearing surfaces are polished and progressively worn toward some optimum. The optimal point depends on the type of food being chewed. Abrasion rate varies according to the physical properties of nutrition: Soft fruits and meat are less abrasive than tough grass fibres or hard seeds. Abrasion velocity increases if the optimum of attrition resistance is exceeded with wear and food composition stays the same.

Much earlier work on mammalian dental anatomy and function concentrated on size and cusp height measurements (Hylander 1975 a, b, Kay & Hylander

1978), as well as on the functional lengths of enamel ridges (Kay 1978). These comparative studies provided the first clues about the correspondence between macromorphological characteristics and dietary strategies within the same mammalian family. Detailed information is also provided by micro-wear analysis of abrasion patterns on wear facets, pits, and scratches occurring through tooth and food contacts (Teaford, 1985, 1988, 1991, 1994; Teaford & Walker, 1982, 1984; Walker 1979; Walker et al. 1978). Such wear patterns document the relative jaw movements and can be used as a source of information for ecoethology (Janis, 1984; Maier, 1984; Teaford, 1991), although micro-wear reflects only the properties of relatively recent meals of an individual.

Additionally, enamel microstructure and thickness allow conclusions about food properties and diet within rough categories to be reached (Beynon & Dean, 1988; Boyde, 1964; Boyde & Martin, 1982, 1987; Gantt, 1983, 1986; Teaford, 2000; Ungar & Teaford, 2001; Shellis & Hiemae, 1986; Shellis et al., 1998).

Analysis of the spatial position, size, and shape of wear facets offers more detailed information about the dietary niche. These facets are described and numbered by Crompton (1971), Crompton & Hiemae (1970), Kay (1973), Kay & Hiemae (1974), and Maier (1977a, b) and applied to phylogenetic analyses by Maier (1977a, b, 1978a, b) and Maier & Schneck (1981). Maier (1978a, b) and Maier & Schneck (1981) underlined the importance of functional associations of wear facets in comparative studies. Thus studies of the chewing cycles of primitive mammals (Crompton & Hiemae, 1970; Hiemae & Kay, 1973; Kay & Hiemae, 1974), for instance, demonstrating that the chewing process is based on almost similar movements, meant that differences in wear pattern could not be explained by distinct chewing behaviours.

Schmidt-Kittler (1984, 1986) characterized the wear pattern of mammalian teeth through a structural parameter, called the D-value, describing the folding of enamel. Based on this and other enamel parameters he interpreted the occlusal pattern of hypsodont molars in a functional and evolutionary context. Kullmer (1997, 1999) further quantified enamel patterns of the occlusal surface with digital two-dimensional (2D) image processing methods. From the correlation of crown height and occlusal surface length with the degree of folding (D-value) in different wear stages, it is possible to model the crown architecture and function (Kullmer, 1999).

This method is limited to relatively flat surfaces, since only a low topographic relief can be reasonably analysed from two dimensional image data sets. Measurement errors occur if we use two-dimensional methods to calculate enamel occlusal surface areas on high topographic relief (Maier, 1980; Janis, 1990). Three-dimensional documentation of tooth surfaces have been accomplished by several research groups. Specific methods of microscopy, such as confocal laser microscopy employed for very small teeth (< 10mm) (Jernvall & Selänne, 1999) and reflex microscopy (Reed, 1997; Strait, 1993), have yielded the first results.

Using Geographic Information System technology developed for the presentation and analysis of landscape surfaces it is feasible to model the crown relief as three-dimensional "landscape", permitting the measurement of cusp volumes and occlusal basins (Zucotti et al., 1998). Zucotti et al. (1998) used an electromagnetic 3D digitising system (Polhemus 3Space) to acquire tooth crown landmark data. However, while this technique is reasonable for relatively large tooth crowns, providing a maximum resolution of about 0.13 mm, the data recording procedure employs a handheld stylus which is consequently relatively time consuming.

Ungar & Williamson (2000) used a laser scanner (RPS 4500 Laserdesign) with a resolution of 0.0254 mm for measuring tooth surfaces. The fixed two-dimensional scan architecture of laser systems consists of a complex opto-mechanical system (Breuckmann, 1993) whose mechanical scanning platform is very limited in object size. Often it is necessary to produce casts of teeth, instead of scanning the original directly because of its attachment to facial skeletal remains.

3D MORPHOMETRICS OF TOOTH CROWNS

The use of 3D datasets for pattern analysis and comparison depends on two major factors. First, the investigator must decide if the method and technique of data acquisition is appropriate in resolution and accuracy to permit the analysis of the region(s) of interest on a specimen. Second, it is important to know the parameter settings of the scanning system in order to estimate the quality of the data. Given suitable imaging conditions, VR reconstructions can be used to extract a wide range of measurements in a virtual environment without touching the original. For instance, the exact location where measurements are taken can be recorded by point-to-point marking in 3D space. Structural parameters, like those for hominid teeth, defined and listed by Wood (1991), can be measured within seconds. As examples, for all specimens examined in this study we provide HOTPAD measurements in Table 1.

Besides distance parameters, Wood (1991) also documented cusp and occlusal surface areas in 2D. To obtain these areas, tracings of photographs were digitised according to the anatomical expertise of the operator, and thus it is very difficult to know the exact margins used for any particular specimen. In virtual reality we can indicate the boundaries on the 3D model and computationally calculate 3D areas. Precise 2D areas are measured after projecting, say, cusp areas of a tooth to a reference plane. Reference planes are necessary if one is to compare data from different teeth controlled for position and orientation. Choosing a proper reference is difficult to define and one should always choose a reference with knowledge of variability and variation.

How shall we choose the best reference plane? The application of landmark methods is limited to precisely defined morphological features. Therefore landmarks cannot help very much in tooth crown examination, since surface morphology changes with wear. We believe that the cervical plane can be used as a reference for the purpose of relief quantification (Figure 5). Wood and Abbott (1983), Wood et al. (1983) and Wood and Uytterschaut (1987) described the calculation of molar base crown, cusp, and premolar base crown and cusp areas, respectively, relative to the cervical plane in 2D, as noted above. In this method, area measures relative to the cervical plane were taken from teeth by means of photographic imaging of the occlusal surface whilst the cervical margin lay perpendicular to the optical axis.

Table 1 - List of structural parameters of tooth crowns provided in HOTPAD, e.g. for molars.

Structural Parameters of Molars (M)

linear measurements 2D [mm] (after Wood (1991): № 313-316; including qualitative parameters № 327, 331-340)		**surface elevation model** [mm, % (deviation), x, y, z] (surface relief to occlusal plane)	
cusp areas 3D + 2D [mm²] (2D: after Wood (1991): № 319-327, 328+329) **3D index** (3D area/2D area)		**best fit cylinder radius & orientation** [mm, % (deviation), x, y, z] (surface models)	
strike and dip cusp sloping, wear facets [°] (stereoplot)		**best fit circle radius & orientation** [mm, % (deviation), x, y, z] (surface models)	
circularity index (2D surface area / max. bounding circle area)		**best fit sphere radius & orientation** [mm,% (deviation), x, y, z] (fitting in surface model, cusp areas, wear facets)	
height index (max. breadth/ max. height) (surface model, cusps)		**best fit plane orientation** [x, y, z] (fitting in cusp areas, wear facets)	

In our study the cervical plane is computationally determined by means of the least squares best fit method. Interactively the cervical margin is marked on the VR model by drawing a closed curve. Afterwards the plane is fitted through all points in 0.5mm distance to curve. The plane is then moved apically toward the deepest point of the fissure pattern. The reference plane cuts the VR model in this position. Further calculations on the occlusal surface are limited to the crown area above the plane which includes the complete occlusal surface. Orientation in the mesio-distal direction remains complicated however. To make orientation measures, such as the strike and dip of cusp slopes, we use the tangents method (Figure 5). In occlusal view we compute two tangents along the lingual and buccal margins. The midpoint between these sub-parallel

reference plane

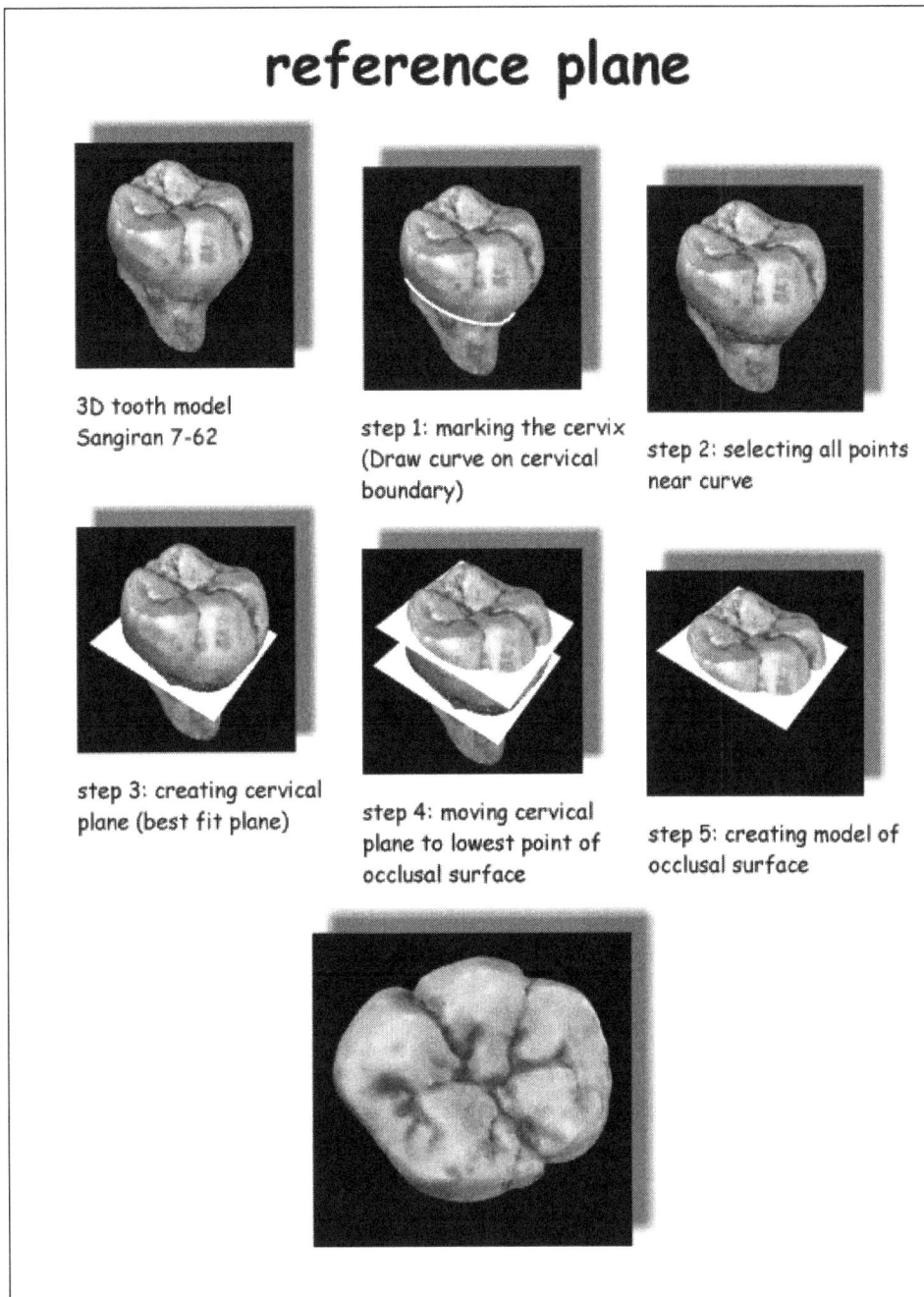

3D tooth model
Sangiran 7-62

step 1: marking the cervix
(Draw curve on cervical
boundary)

step 2: selecting all points
near curve

step 3: creating cervical
plane (best fit plane)

step 4: moving cervical
plane to lowest point of
occlusal surface

step 5: creating model of
occlusal surface

Figure 5 - Creating the cervical plane on a virtual model as reference for analyses of
occlusal surfaces using the PolyWorks software package.
Step 1: Draw curve on cervix interactively. Step 2: Selecting all points near curve (distance
0.5mm). Step 3: Creating best fit plane through points. Step 4: Align coordinate system (x,y)
to cervical plane and move plane towards lowest point of the fissure pattern. Step 5: Cut
model with plane and reconstruct occlusal surface model for morphometric analysis.

tangents is located and a precise mesio-distal longitudinal axis chord is formed, similar to the 3D method outlined by Bromage et al. (1995).

Slope is calculated by fitting a plane into the marked cusp slope. Strike and dip of the fitted plane is extracted by common geological methods. The results can be visualized in a stereoplot diagram (Figure 6). Each great circle and pole reflects the overall orientation of slope of a defined area on the tooth surface. The steeper the relief the closer is the pattern of great circles located around the centre of the diagram. Each occlusal surface will be represented by its individual stereoplot

pattern. Additionally we can use stereoplot diagrams to quantify the wear facets (Figure 7) on the occlusal surface to determine wear stages and to analyse and document shearing and grinding capacities of teeth (Kullmer et al., in prep).

DISCUSSION AND PERSPECTIVES OF 3D TOPOMETRY

Summarizing the main features of optical 3D topometry, we may point out its potential as follows:

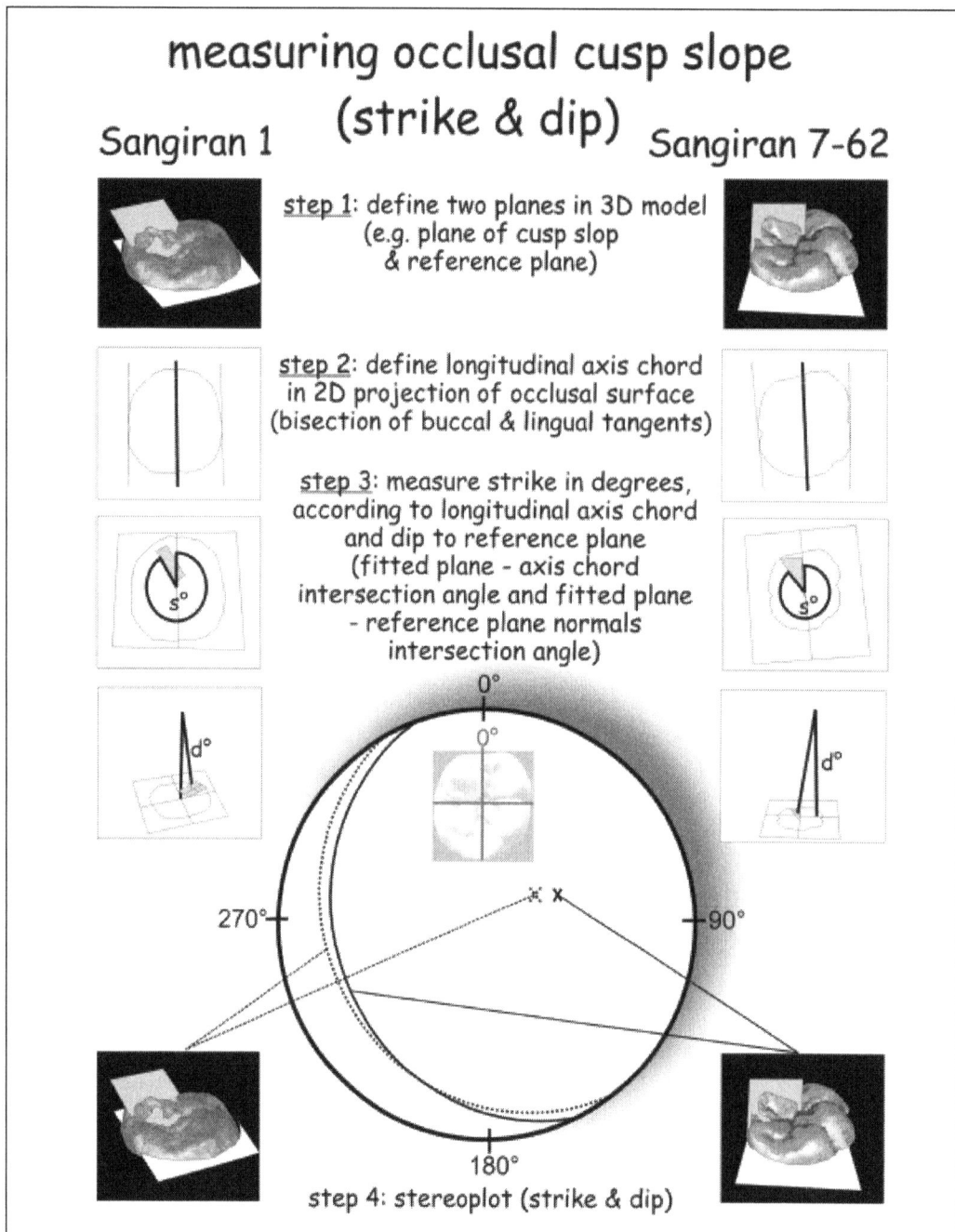

Figure 6 - Measuring cusp sloping, including strike and dip in degrees. Step 1: Defining cervical plane as reference and plane of cusp sloping, using the best fit least square method. Step 2: Defining mesio-distal longitudinal axis chord in 2D projection of occlusal surface boundary, using buccal and lingual tangents. Step 3: Measuring strike in degrees according to longitudinal axis chord and dip to reference plane orientation. Step 4: Presentation of strike and dip in a stereoplot diagram. Position of great circles and poles in the lower hemisphere projection reflecting average orientation of occlusal cusp slope.

- Fast 3D digitising of object surfaces of modern and fossil originals.
- High accuracy
- High resolution
- Large dataset
- Non-contact data acquisition
- High flexibility in object size
- Portable system

However, not all objects are suitable for this method, depending on the particular question. The choice of the measuring volume is an important factor, since the maximum resolution for the VR model is restricted. It is important to know in advance on which details the investigation takes place before scanning the object. The measuring of very complex objects can lead to problems during scanning, resulting in missing data because of shadows and deep cavities on the surface. Very dark or even black specimens cannot be measured without powdering to uniformly lighten the surface. Rapidly and intensely changing lights should be avoided, since otherwise no exact calculation of light intensities is possible and the measuring sequence has to be repeated. However,

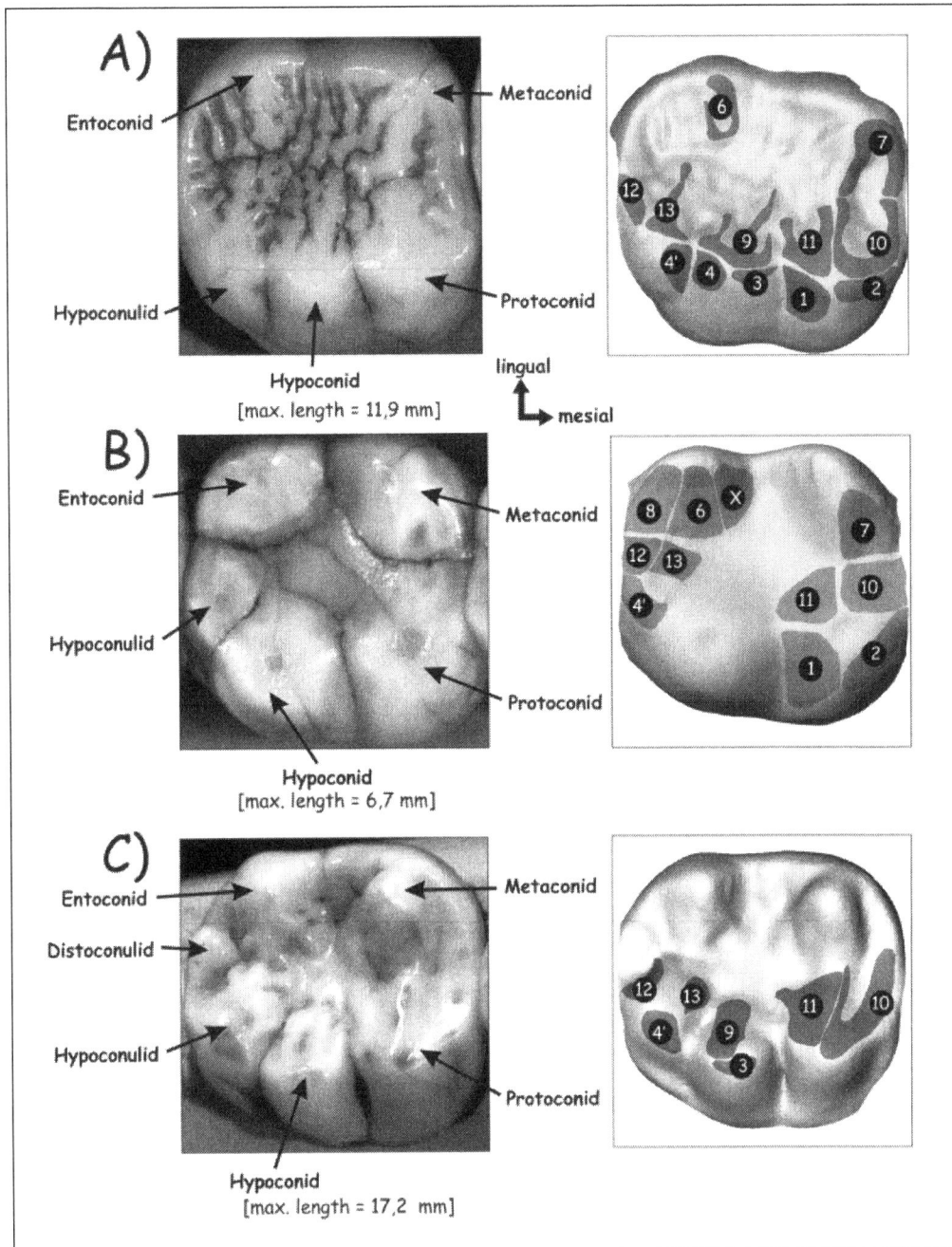

Figure 7 - A) (left) Digital image of lower right M2 of an orang-utan (*Pongo pygmaeus*), (right) virtual 3D model with marked wear facets (numbering after Crompton, 1971, Kay 1973 and Maier 1977). Facets 8 and 5 are not identifiable, because of crenulated enamel.
B) (left) Digital image of lower right M2 of a gibbon (*Hylobates hoolock*), (right) virtual 3D model with marked wear facets (numbering after Crompton, 1971, Kay 1973 and Maier 1977). Facets 3,4 and 9 are missing, because of rounded, dome-shaped hypoconid. An additional facet (x) is observed, close to entoconid. Facet 5 is missing. C) (left) Digital image of lower right M2 of a gorilla (*Gorilla gorilla*), (right) virtual 3D model with marked wear facets (numbering after Crompton, 1971, Kay, 1973 and Maier, 1977). Facets 1, 2, 4, 5, 6, 7 and 8 are not identifiable on the gorilla M2.

some of the limitations of the system can be influenced positively by individual system parameter settings, depending on object properties as noted above and the scan environment.

The optical 3D topometry possesses some important advantages for the use in palaeontology and archaeology compared to alternative measuring techniques, such as high flexibility in object size and portability. High resolution 3D surface data extends morphometric possibilities enormously.

The fast measuring of surface structure in space replaces time-consuming measuring procedures and provides more accurate and precise data. Presently 3D scanning techniques are tested in many fields of palaeontology. Digital 3D data are used for structural analysis, virtual reconstruction, animation, presentation and archives. Remarkable are the didactic possibilities of computer animation and reconstruction. One can magnify the digital model by the computer to reproduce an enlarged replica via stereo-lithography or modern 3D printing devices. HOTPAD is the first attempt to document a hominid collection with high-resolution VR models of specimens to demonstrate the potential of 3D digital image processing. The current project should be understood to be an exploration into the universe of Virtual Reality in palaeoanthropology. In the near future digital techniques like optical topometry will join with computer tomography and confocal microscopy to illuminate new 3D aspects in palaeontological research.

SUMMARY

HOTPAD (Hominid tooth pattern database) is a high quality digital archive of virtual models. It can be used as a catalogue to provide colleagues and interested parties all over the world the chance to estimate the value of specimens in an archived collection for their research.

The use of digital three-dimensional (3D) image processing systems permits the extraction of structural parameters within seconds and, together with modern scanning techniques, opens an entirely new perspective for morphological analysis in palaeobiology.

In this contribution we document a novel 3D technique for the evaluation and quantification of hominid specimens collected by G.H.R von Koenigswald in Indonesia. The resultant database will be used for more extensive studies on this important assemblage.

The optical topometry system used by us (Breuckmann) consists of a sensor with a power supply and a notebook computer (PIII 650Mhz, 20Gb HD). The sensor is connected to a tripod and can be rotated in all directions. The whole system is portable and fits into one small box. It can be transported to museums for digitising collections with a resolution up to 50µm.

After digitising, the calculated Virtual Reality (VR) model can be analysed in absence of the original. The VR models are placed in a tooth pattern database called HOTPAD which can be viewed from a normal Internet browser. One can use HOTPAD as an information centre on hominid tooth patterns or just to look at specimens to assess their completeness. But further, information about the 3D scanning, measurement lists, and comparisons of results of the fossil material are also available. The first version of HOTPAD contains molar material from Sangiran. As an add-on we include the fossil skull fragments of this collection. VR models in different resolutions are available for downloading, including measurements for the purpose of study and analysis. HOTPAD contains conventional measurements, like distance parameters based on the definitions of Wood (1991) and complex 3D measurements, including relative parameters such as circularity and relief indices. 2D and 3D areas of cusps and occlusal surface are also provided. Furthermore, industrial 3D software, normally used for the inspection of product parts, allows us to view the exact elevation of each calculated 3D point. HOTPAD should be understood as a contribution to enlighten the possibilities of using modern 3D techniques to judge queries in palaeontological and palaeoanthropological research.

Acknowledgements

We would like to thank Dr Bertrand Mafart and Hervé Delingette for the invitation to this inspiring colloquium. Many thanks to the UISPP committee 2001, which accepts to publish our colloquium communication. This research is enabled by the Forschungsinstitut Senckenberg in Frankfurt am Main. We are grateful to Bernd Herkner for his critical and innovative comments on HOTPAD.

Authors' addresses:

Ottmar KULLMER, Mathias HUCK, Kerstin ENGEL, Friedemann SCHRENK

Forschungsinstitut und Naturmuseum Senckenberg, Paläoanthropologie

Senckenberganlage 25, 60325 Frankfurt am Main, Germany

Timothy BROMAGE

Department of Anthropology

Hunter College, CUNY

695 Park. Ave., New York, NY 10021, USA

Correspondance to: Ottmar Kullmer, okullmer@sng.uni-frankfurt.de

BIBLIOGRAPHY

BEYNON, A.D. & Dean, M.C., 1988, Distinct dental development patterns in early fossil hominids. - *Nature* 335, 509-514.

BEYNON, A.D. & WOOD, B.A., 1986, Variations in Enamel Thickness and Structure in East African Hominids. - *Am. J. Phys. Anthrop.* 70, 177-193.

BOYDE, A., 1964, *The structure and development of mammalian enamel.* - Ph.D. Thesis. Univ. of London.

BOYDE, A. & MARTIN, L.B., 1982, Enamel microstructure determination in hominoid and cercopithecid primates. - *Anat. Embryol.* 165, 193-212.

BOYDE, A. & MARTIN, L.B., 1987, Tandem Scanning Reflected Light Microscopy of Primate Enamel. - *Scanning Microscopy 1*, 1935-1948.

BREUCKMANN, B., 1993, *Bildverarbeitung und optische Meßtechnik in der industriellen Praxis.* - Franzis, München, 1993, S. 1-315.

BROMAGE, T.G., SCHRENK, F. & ZONNEVELD, F.W., 1995, Palaeoanthropology of the Malawi Rift: An early hominid mandible from the Chiwondo Beds, northern Malawi. *Journal of Human Evolution* 28, 71-108.

CROMPTON, A.W., 1971, The origin of the tribosphenic molar. - *Zool. J. Linn. Soc.* 50 (Suppl.), 65-87.

CROMPTON, A.W. & HIIEMAE, K.A., 1970, Molar occlusion and mandibular movements during occlusion in the American opossum, Didelphis marsupialis L. - *Zool. J. Linn. Soc.* 49, 21-47.

GANTT, D.G., 1983, *The Enamel of Neogene Hominoids. Structural and Phyletic Implications.* - In *New Interpretations of Ape and Human Ancestry*, edited by R.L. Ciochon, & R.S. Corruccini, 1983, New York/London: Plenum Press, 249-298.

GANTT, D.G., 1986, Enamel Thickness and Ultrastructure in Hominoids: With Reference to Form, Function and Phylogeny. - In *Comparative Primate Biology*, edited by D.R. Swindler & J. Erwin, Vol. 1: Systematics, evolution and anatomy. 1986, New York: Alan R. Liss, 453-475.

GRINE, F. E. & FRANZEN, J. L., 1994, Fossil Hominid Teeth from the Sangiran Dome (Java, Indonesia). *Cour. Forschungsinstitut Senckenberg*, 171: 75-103, 1 Tab., 4 Pls.

HIIEMAE, K.M. & KAY, R.F., 1973, Evolutionary Trends in the Dynamics of Primate Mastication. - In *Symp IVth Int. Congr. Primat., vol. 3: Craniofacial Biology of Primates.* 1973, Basel: Karger, p.28-64.

HYLANDER, W.L., 1975a, Incisor Size and Diet in Anthropoids with Special Reference to Cercopithecidae. - *Science* 189, 1095-1098.

HYLANDER, W.L., 1975b, Incisor Size and Diet in Cercopithecidae. - *Am. J. Phys. Anthrop.* 42, 309.

JANIS, C.M. , 1984, Predictions of primate diets from molar wear pattern. In *Food Acquisition and Processing in Primates* edited by Chivers, B.A. et al. 1984, Plenum Publishing Corporation, p.331-340.

JANIS, C.M., 1990, The correlation between diet and dental wear in herbivorous mammals, and its relationship to the determination of diets of extinct species. In *Evolutionary Paleobiology of Behaviour and Coevolution* edited by A.J. Boucot. 1990, Amsterdam: Elsevier Science Publishers B.V., p.241-259.

JERNVALL, L. & SELÄNNE, L. (1999). Laser confocal microscopy and geographic information systems in the study of dental morphology.- *Palaeontologica Electronica*, Vol. 2(1): http://www-odp.tamu.edu/paleo/1999_1/confocal/issue1_99.htm.

KAY, R.F., 1978, Molar Structure and Diet in Extant Cercopithecidae. - In *Development, Function and Evolution of Teeth*, edited by P. Butler & K. Joysey. 1978, New York/London: Academic Press, p.309-339.

KAY, R.F. & HIIEMAE, K.M., 1974, Jaw movement and tooth use in recent and fossil primates. - *Am. J. Phys. Anthrop.* 40, 227-256.

KAY, R.F. & HYLANDER, W.L., 1978, The Dental Structure of Mammalian Folivores with Special Reference to Primates and Phalangeroidea (Marsupialia). - In *The Ecology of Arboreal Folivores* edited by G.G. Montgomery. 1978, Washington: Smithonian Institute Press, p.173-191.

KOENIGSWALD, G.H.R. von, 1940, Neue Pithecanthropus-Funde 1936-1938. Ein Beitrag zur Kenntnis der Praehominiden.- *Wet. Meded.*, 28: 1-232.

KOENIGSWALD, G.H.R. von, 1950, Fossil hominids of the lower Pleistocene of Java.- *Proc. Int. Geol. Congr. Gr. Britain* 1948, Part 9, sect. H: 59-61.

KOENIGSWALD, G.H.R. von, 1954, The Australopithecinae and Pithecanthropus. I – III. – *Proc. kon. nederl. Akad. Wet.* (B) 56: 403-413; (B) 56: 427-438; (B) 57: 85-91.

KULLMER, O. (1997). *Die Evolution der Suiden im Plio-Pleistozän Afrikas und ihre biostratigraphische, paläobiogeographische und paläoökologische Bedeutung.* - Dissertation im Fachbereich Geowissenschaften der Johannes Gutenberg-Universität, Mainz, pp.143.

KULLMER, O., 1999, Evolution of African Plio-Pleistocene Suids (Artiodactyla: Suidae) Based on Tooth Pattern Analysis. - *Kaupia, Darmstädter Beiträge zur Naturgeschichte* 9, 1-34.

KULLMER, O., SCHRENK, F., DÖRRHÖFER, B., in press, High resolution 3d-image analysis of ape, hominid and human footprints. *Cour. Forschungsinstitut Senckenberg*.

KULLMER, O., ENGEL, K. & SCHRENK, F., in prep, Moderne dreidimensionale Vermessungstechniken erweitern paläontologische Forschung - Virtuelle Zahnmodelle ermöglichen die Analyse komplexer Zahnoberflächen.

MAIER, W., 1977a, Die Evolution der bilophodonten Molaren der Cercopithecoidea. Eine funktionsmorphologische Untersuchung. - *Z. Morph. Anthrop.* 68, 26-56.

MAIER, W., 1977b, Die bilophodonten Molaren der Indriidae (Primates) - ein evolutionsmorphologischer Modellfall. - *Z. Morph. Anthrop.* 68, 307-344.

MAIER, W., 1978a, Zur Evolution des Säugetiergebisses - Typologische und konstruktionsmorphologische Erklärungen. - *Natur und Museum* 108(10), 288-300.

MAIER, W., 1978b, Die Evolution der tribosphenischen Säugetiermolaren. - *Sonderbd. Naturwiss. Verh.* 3, 41-60.

MAIER, W., 1980, Konstruktionsmorphologische Untersuchungen am Gebiß der rezenten Prosimiae (Primates). - *Abhandlungen der Senckenbergischen Naturforschenden Gesellschaft* 538, 1-158.

MAIER, W., 1984, Tooth Morphology and Dietary Specialization. - In *Food Acquisition and Processing in Primates* edited by D.J. Chivers. 1984, Plenum Publishing Corporation, p.303-330.

MAIER, W. & SCHNECK, G., 1981, Konstruktionsmorphologische Untersuchungen am Gebiß der hominoiden Primaten. - *Z. Morph. Anthrop.* 72, 127-169.

REED, D.N.O., 1997, Contour mapping as a new method for interpreting diet from tooth morphology. - *Am. J. Phys. Anthrop.*, Suppl. 24, 194.

SCHMIDT-KITTLER, N., 1984, Pattern analysis of occlusal surfaces in hypsodont herbivores and its bearing on morphofunctional studies. - *Koninkl. Nederl. Acad. Wetensch., Proceedings* B 87 (4), 453-480.

SCHMIDT-KITTLER, N., 1986, Evaluation of occlusal patterns of hypsodont rodent dentitions by shape parameters. - *N. Jb. Geol. Paläont.* Abh. 173(1), 75-98.

SHELLIS, R.P. & HIIEMAE, K.M., 1986, Distribution of Enamel on the Incisors of Old World Monkeys. - *Am. J. Phys. Anthrop.* 71, 103-113.

SHELLIS, R.P., BEYNON, A.D., REID, D.J. & HIIEMAE, K.M., 1998, Variation in molar enamel thickness among primates. - *J. Hum. Evol.* 35(4/5), 507-522.

STRAIT, S.G., 1993, Differences in occlusal morphology and molar size in frugivores and faunivores. *J. Hum. Evol.* 25(6), 471-482.

TEAFORD, M.F., 1985, Molar microwear and diet in the genus Cebus. - *Am. J. Phys. Anthrop.* 66, 363-370.

TEAFORD, M.F., 1988, A review of dental microwear and diet in modern mammals. - *Scanning Microscopy 2*, 1149-1166.

TEAFORD, M.F., 1991, Dental microwear: What can it tell us about diet and dental function? - In *Advances in Dental Anthropology* edited by M. A. Kelley & C. S. Larsen. 1991, New York: Wiley-Liss, p.341-356.

TEAFORD, M.F., 1994, Dental microwear and dental function. - *Evol. Anthrop.* 3, 17-30.

TEAFORD, M.F., 2000, Primate dental functional morphology revisited. – In *Development, Function and Evolution of Teeth* edited by M. F. Teaford, M. M. Smith & M. W. J. Ferguson. pp. 314, Cambridge University Press.

TEAFORD, M.F. & WALKER, A., 1982, Dental microwear: mechanisms and interpretations. - *Am. J. Phys. Anthrop.* 62, 236.

TEAFORD, M.F. & WALKER, A., 1984, Quantitative differences in dental microwear between primate species with different diets and a comment on the presumed diet of Sivapithecus. - *Am. J. Phys. Anthrop*. 64, 191-200.

UNGAR, P.S. & WILLIAMSON, M., 2000, Exploring the effects of tooth wear on functional morphology: A preliminary study using dental topographic analysis. - *Palaeontologica Electronica*, Vol. 3(1): http://www-odp.tamu.edu/paleo/2000_1/gorilla/ issue1_00.htm.

UNGAR, P.S. & TEAFORD, M., 2001, The Dietary Split Between Apes and the Earliest Human Ancesters.- In *Humanity from African Naissance to Coming Millenia,* edited by P. V. Tobias, M. A. Raath, J. Moggi-Cecchi, G. A. Doyle. pp. 409, Firenze University Press.

WALKER, A., 1979, S.E.M. Analysis of Microwear and its correlation with dietary patterns. - *Am. J. Phys. Anthrop*. 50, 489.

WALKER, A., HOECK, H.N. & PEREZ, L., 1978, Microwear of mammalian teeth as an indicator of diet. *Science* 201, 908-910.

WOOD, B., 1991, *Hominid Cranial Remains. Koobi Fora Research Project. 4*. Oxford: Clarendon.

WOOD, B.A., 1993, Early Homo: How many species? In *Species, Species Concepts, and primate Evolution*, edited by W.H Kimbel. & L.B. Martin. pp. 485-522, New York: Plenum Press.

WOOD, B.A. & ABBOTT, S.A., 1983, Analysis of the dental morphology of Plio-Pleistocene hominids: I. Mandibular molars: crown area measurements and morphological traits. *J. Anat*. 136, 197-219.

WOOD, B.A. & UYTTERSCHAUT, H., 1987 Analysis of the dental morphology of Plio-Pleistocene hominids: III.Mandibular premolar crowns. *J. Anat*. 154, 121-156.

ZUCCOTTI, L.F., WILLIAMSON, M.D., LIMP, W.F., UNGAR, P.S., 1998, Modeling primate occlusal topography using geographic information systems technology. - *Am. J. Phys. Anthrop*. 107, 137-142.

SIMULATION AND 3D-LASER-SCANNING OF DENTAL ABRASION

Irene Luise GÜGEL & Karl-Heinz KUNZELMANN

Résumé: La perte d'émail dentaire sous l'effet des particules abrasives contenues dans l'alimentation peut avoir des conséquences importantes pour la santé et le bien–être des sujets. Les connaissances actuelles sur les processus de formation de cette abrasion sont encore limitées. Dans le cadre de nos recherches, une étude expérimentale a été conduite pour mieux simuler et quantifier l'usure dentaire obtenue avec des céréales pour des modes de broyages différents. Des échantillons d'émail de molaires humaines extraites ont été utilisées comme matériel expérimental dans un simulateur de mastication. Les surfaces de l'émail ont été scannérisées avant et après cette mastication expérimentale, avec un Laserscanner Pro 3D. L'importance de l'usure totale a été quantifiée à partir des données tridimensionnelles par superposition des images numériques pré et post procédure grâce à un algorythme des moindres -carrés sans utilisation de points de référence. L'« image de la différence » était obtenue par soustraction digitalisée des deux images numériques 3D superposées.

Abstract: The loss of dental enamel caused by abrasive particles in the diet can have serious consequences for human health and well-being. With regard to the formation process of abrasion, current knowledge is still limited. In the course of our research, an experimental approach served for a better simulation and quantification of dental abrasion, depending on different cereal species and milling techniques. Enamel samples of impacted human molars were used as agonists in a chewing simulator. The enamel surfaces were scanned before and after the chewing simulation by a Laserscanner Pro 3D. Absolute abrasion was determined by this 3D-data set, by numerically superimposing the baseline and follow-up image with a least-squares matching algorithm without reference points. The "difference image" was calculated by digital substraction of the superimposed 3D-data sets of loss.

INTRODUCTION

The pattern of tooth wear is related to cultural, dietary, occupational, and geographical factors. Tooth wear analysis continues to play an important role in the understanding of the biological consequences of subsistence change, which is but one component in this complex analysis. The abrasiveness of food is a key determinant in the rate of physiological dental wear in humans (Molleson et al., 1993, Molnar 1971, 1972, Powell 1985, Rose et al. 1985, Rose & Ungar, 1998, Smith, 1984, Teaford & Lytle, 1996). With increasingly refined food processing methods through time, the rate of physiological dental wear on human teeth has changed markedly.

Many investigators share the opinion that a certain degree of dental wear is beneficial for dental health. This was demonstrated by the elimination of cuspal interferences to excursive movements and the reduction of caries and periodontal diseases by the removal of stagnating areas (Ainamo, 1972, Berry and Poole, 1976). A lack of occlusal and approximal dental wear may also lead to crowding, rotation and overlapping of anterior teeth (Lombardi 1982, Wolpoff, 1971).

The analysis of abrasion on teeth of pre-modern human populations led to important data about the degree and patterns of abrasion (for review: Vötter, 1973, Larsen, 1997). However the underlying causes of different abrasion rates and patterns, and also the wear formation process as such are far from being clear. The aim of the following study was to further investigate the process of abrasion on human enamel by use of a chewing-device for a simulation of abrasion, and subsequent analysis of enamel loss by 3D-Laser-Scanning (Kunzelmann, 1997).

MATERIALS AND METHODS

In vitro simulation of abrasion

In vitro enamel abrasion on human enamel samples was performed with the Academic Center for Dentistry Amsterdam (ACTA) device (DeGee et al., 1986, Gügel et al., 2001). The ACTA chewing simulator consists of both a sample wheel and an antagonistic wheel with different diameters (Gügel et al., 2001). 20 chambers of the sample wheel were filled with dental filling material, and a cavity was cut into 10 separated chambers each. Ten prepared human enamel samples of impacted teeth were cemented into these cavities. The "occlusal" part of the enamel insert was lifted up to 850 mm maximum above, while the "cervical" part was left at the polished surface. SEM investigations after the simulation process did not reveal any significant signs of fatigue, indicative of a possible rise of pressure exceeding physiological biting forces. The chambers without enamel served as internal controls. During the experiments, the wheels were completely covered by the diet slurry. The wheels rotated against each other with different velocities, so that their sliding motion simulates the lateral sliding of antagonistic teeth in the jaws during the crushing phase of the chewing cycle. The antagonistic wheel works with a constant load of 15 N to simulate physiological biting forces. The slurries included cereal species which were expected to induce rapid (millet), moderate (spelt and wheat) and low (spelt) enamel loss according to their differing abrasiveness. The slurries contained amounts of phytoliths, amorphous silica particles, the occurance of which is largely plant-specific and contaminating particles which are one factor for abrasion caused by diet (Teaford and Lytle, 1996, Gügel et al., 2001). Wheat and spelt were ground on historical sandstone, millet was ground with its silica-rich husks, and a negative control without debris consisted of modern prepared spelt. To simulate the course of abrasion on human enamel in the ACTA-device, first experiments were performed with slurry consisting of spelt, which was roughly ground in a modern mill with korund-ceramics. High resolution replicas were taken as baseline before the chewing simulation, and follow-up after every 50 000 cycles of the simulation until 200 000 cycles. It was expected that this procedure would induce a loss of small quantities of enamel, which could already be detected stepwise after every 50 000 cycles by the 3D-Laser-Scanning method.

3D-Laser-Scanning of experimentally abraded enamel surfaces

The measurement on the abraded areas of the samples was achieved by scanning high resolution replicas of the enamel surface (Beynon 1987) by the 3D-Laser Scanner (Mehl et al., 1997). The preparation of replicas was necessary to overcome translucency of natural enamel. A light line created by a laser diode enables the measurement of about 512 surface points within 40 msec (that is 500x faster compared to a profilometric method) which are projected onto a CCD-chip under a defined triangulation angle. The information of height for any surface point is encoded within the lateral displacement of the light line. By shifting the enamel sample on a stepper motor along the y-axis, the entire 3D-surface can be measured line by line (Figure 1a, b). The CCD-image is stored in a frame-grabber. To measure small surface changes, the two digital images (from the baseline, Figure 1a, and follow-up, Figure 1b, scannings) were superimposed (Figure 1c).

The software (Mehl et al., 1997) was developed as a reference-free 3D-superimposing algorithm. The fitting region of reference points in corresponding images can now be defined

by the operator. Areas of wear and other differences between baseline and follow-up model can thus be excluded. The matching process is considered successful when the difference values in known identical regions show a "noisy" pattern with its standard deviation close to the sensor accuracy. The loss of enamel was finally made visible by presentation in counter-colours. The mean loss of height and volume was calculated for each wear region on the basis of the difference-image. The results obtained were compared with the results achieved by the much more timeconsuming and therefore limited method of profilometry with a perthometer on the identical samples.

RESULTS

Absolute abrasion loss

Cereal species with different phytolith contents and contaminating particles from the grinding process show differences in their abrasiveness (Gügel et al., 2001). Wheat and spelt ground on historical sandstone show abrasion values which fall between a negative control without debris consisting of modern prepared spelt and the highly abrasive millet, which was ground with its silica-rich husks in both scanning procedures (Table 1). According to nonparametrical statistical analysis with SPSS 10.0 (Kruskal-Wallis-Test) no significant differences in the ranking of abrasion due to the scanning method occured (Table 2a).

The differences (U-Mann Whitney, >2 independent samplings) in total abrasion between the control and all the cereal species tested are significant on the 5% level for all parameters analysed by the 3D-Laser-scanning and the

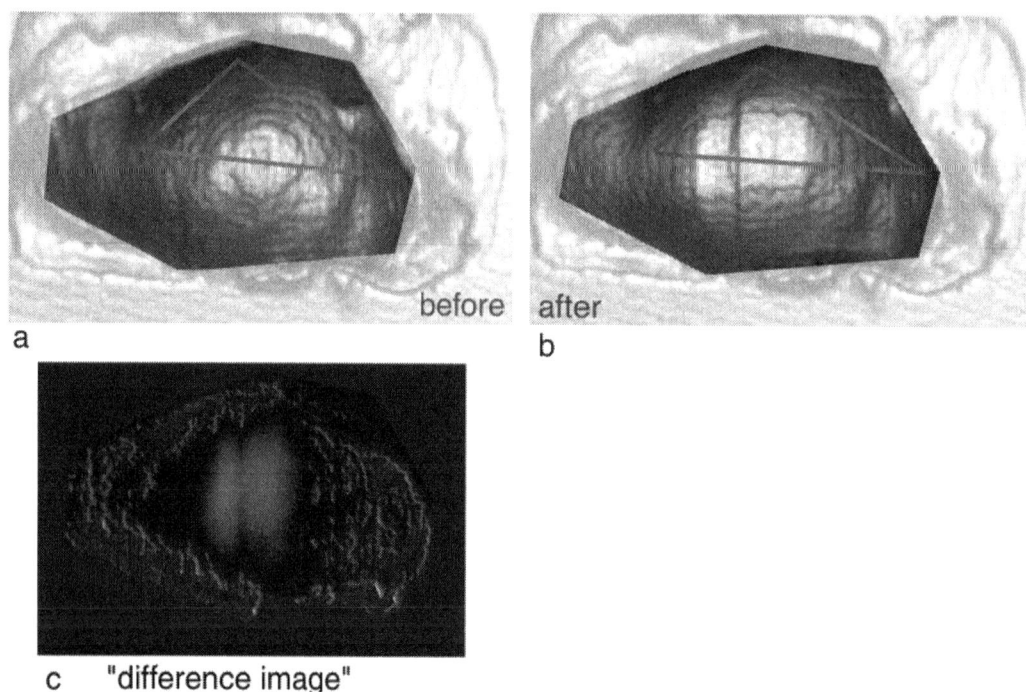

a: before
b: after

c "difference image"

Figure 1 - 3D-Laser-Scanning of experimentally abraded enamel surfaces:
a: before abrasion, b: after abrasion, c: difference image

Table 1 - Median enamel abrasion after 200,000 cycles of chewing simulation with ACTA device using two different scanning procedures

		Spelt (b)[2]	Spelt (m)[2]	Wheat (m)[2]	Millet (h)[2]
% ash residue per dry weight[1]		< 0.01	0.20	n.d.[3]	1.46
3D-Laserscanning	Median abrasion loss (μm) ± 1sd[4]	-7.6 (1,5)	-21.8 (4.1)	-22.1 (6.7)	-33.2 (6.2)
Profilometry	Median abrasion loss (μm) ± 1sd[4]	-8.3 (3.1)	-24.6 (4.2)	-17.8 (2.3)	-27.1 (5.4)
	N (samples)	4	4	4	4

[1] After both wet and dry ashing, consisting mainly of silica. (cf. Gügel et al. 2001)

[2] b, bioshop, ground in mill with korund-ceramics; m, ground in historical mill; h, ground in modern householdmill;

[3] Not detected, but usually similar to rye and oat, 0.01 – 0.03; Runge pers. comm

[4] Standard deviation in parenthesis

Table 2a - Nonparametrical test (Kruskal-Wallis-Test) to compare two different scanning procedures by ranking the abrasivity of cereals (median abrasion loss).

Ranking in average	Spelt (b)	Spelt (m)	Wheat (m)	Millet (h)
3D-Laserscanning	28,50	15,25	13,75	4,50
Profilometry	28,50	12,00	19,75	9,75
Level of significance 0,1 %.				

Table 2b - Nonparametrical test (Mann-Whitney-U-Test) to compare the abrasivity of cereals (median abrasion loss).

	3D-Laserscanning			Profilometry		
	Spelt (m)	Wheat (m)	Millet (h)	Spelt (m)	Wheat (m)	Millet (h)
Spelt (b)	0,029*	0,029*	0,029*	0,029*	0,029*	0,029*
Spelt (m)		0,886	0,029*		0,200	0,686
Wheat (m)			0,029*			0,114
Level of significance* 5 %.						

(b), (m), (h) see Table 1.

profilometric method (Table 2b). In contrast the differences between wheat or spelt to the high abrasive millet remained insignificant after the application of the profilometric method, whereas there is a significant difference on the 5% level after the application of the 3D-Laser-Scanning method (Table 2b). Though the homogeneity of the variances between the scanning methods cannot be statistically rejected, there is a higher statistical security for homogeneity for different cereal species in the 3D-Laser-scanning method opposed to the profilometry. The results indicate that the 3D-Laser-Scanning method permits a more reliable analysis of simulated abrasion on human enamel samples in a very short time.

The abrasion process

In addition these results allowed for an experimental approach to the abrasion process. Table 3 and Figure 2a and 2b show the interrelationship between the number of chewing cycles and four different abrasion parameters: the median, the mean, the maximum and the volume of enamel abrasion loss of every individual enamel sample.

The results of 150 000 cycles are missing because of technical problems. Though the mean tends to be slightly higher than the median, indicative of an asymmetrical distribution of abrasion on the enamel area, all values show a continuing decrease of enamel height and volume with increasing chewing-cycles, as expected. The linear regression of enamel loss of height for every individual enamel sample shows a high correlation between abrasion value and chewing cycles, 3 out of five samples correlate with a coefficient better than -0,900. The analysis of the median abrasion loss of the five data sets leads to a linear regression with a slope of –0,0001 and a constant of –4,567. The linear regression of the mean and the median enamel loss is highly significant on the level of 0,1 % and the loss of volume on the level of 1 %. The

Table 3 - Quantitative enamel loss during the abrasion process.

sample	cycles	maximum abrasion loss (µm)	mean abrasion loss (µm)	median abrasion loss (µm)	volume of abrasion loss (µm^3)
1	50000	-84,6	-17,7	-17,1	-5,07E+07
	100000	-103,3	-22,6	-22,9	-7,64E+07
	200000	-126,2	-37,5	-34,7	-1,40E+08
				r = 0,919	
3	50000	-332,7	-27,6	-20,3	-1,53E+08
	100000	-367,0	-38,5	-28,3	-2,57E+08
	200000	-387,7	-52,6	-27,1	-3,75E+08
				r = 0,618	
4	50000	-61,3	-13,1	-13,2	-6,72E+07
	100000	-89,7	-14,1	-14,9	-1,02E+08
	200000	-105,7	-22,0	-21,1	-1,13E+08
				r = 0,824	
5	50000	n.d.	-47,4	-25,1	-3,02E+08
	100000	n.d.	-57,9	-35,3	-3,83E+08
	200000	n.d.	-84,3	-72,7	-4,86E+08
				r = 0,986	
6	50000	-52,0	-6,3	-5,1	-2,04E+07
	100000	-81,8	-10,6	-11,3	-5,07E+07
	200000	-92,9	-18,7	-14,6	-7,64E+07
				r = 0,918	
linear regression	50000	-94,5	-14,2	-14,6	-6,25E+07
	100000	-139,5	-24,2	-24,6	-1,05E+08
	200000	-229,5	-44,2	-44,6	-1,90E+08
r (correlation)		0,214	0,559	0,598	0,401
slope		-0,0008	-0,0002	-0,0001	-849,09
constant		-49,534	-4,223	-4,567	-2,00E+07
Significance on the level of 1%*, 0,1%**		0,071	0,001**	0,000**	0,008*

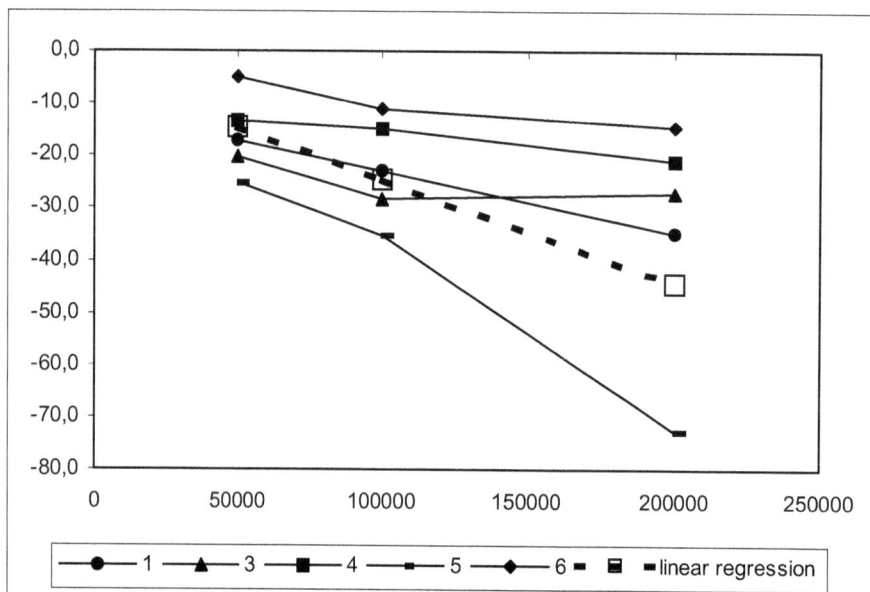

Figure 2a - Median enamel height loss (µm) during the abrasion process (n = 5).

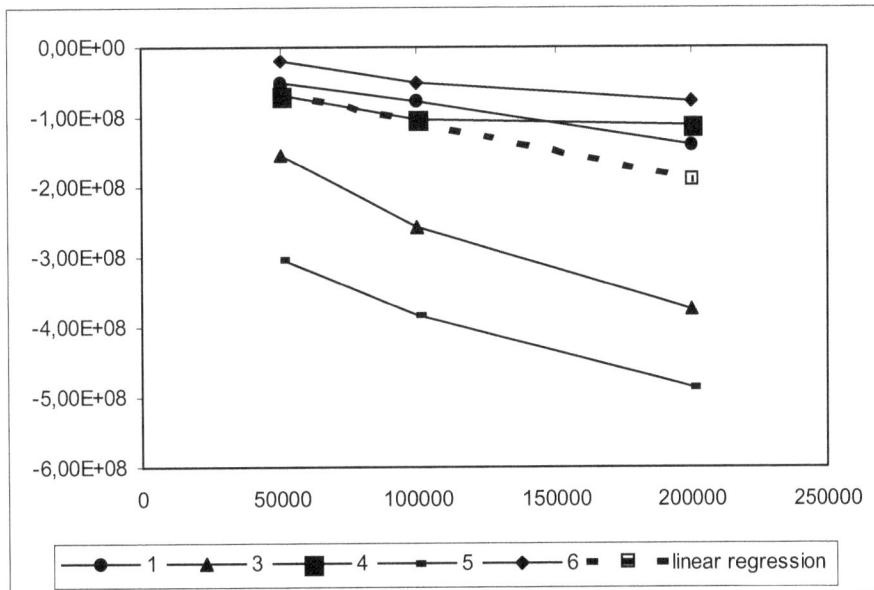

Figure 2 b - Loss of enamel volume (µm³) during the abrasion process (n = 5).

linear regression of maximal enamel height loss is not significant. The low correlation coefficient of –0,598 also indicates the high variability of individual resistance of enamel towards the abrasive effect of diets.

SUMMARY

Our first results on the experimental investigation of the effect of diets on the rate of abrasion are encouraging. We were able to show that the 3D-Laser-Scanning method can replace the profilometric scanning of abraded surfaces, and that a continuous loss of small amount of enamel is detectable on high resolution replicas of enamel samples with experimentally induced wear. The necessity of replica preparation, because of the translucency of natural enamel, does not prevent the detection of little abrasion which is necessary for the discrimination between both different rates of abrasion and different abrasivity of diets. The preparation of high resolution replicas permits at the same time insights into the experimental formation process of dental microwear. The time consuming procedure of profilometry (which needs nearly 16 h for one sample wheel) did not permit the collection of this data set. Further experiments are needed for the establishment of the 3D-Laser-scanning method on abraded areas resulting from different abrasive foodstuffs and debris.

With the ACTA-simulation and the analytical procedure presented in this paper we introduce the application of a methodological spectrum used in modern dentistry suitable for getting access to experimentally produced data on the abrasivity and, in particular, the rate of abrasion in relation to different diets and subsistence strategies on human enamel, independent of the age of the tooth or dental disease. The results of further experiments with a different chewing device to better simulate physiological conditions are in preparation and will be published soon.

Acknowledgements

Thanks to Sybille Friedel, Central Institute of Medicinal Techniques, Garching, Germany and Wolfram Gloger, Policlinics of Dentistry and Parodontology, Munich, Germany for their cooperation, and to the Deutsche Forschungsgemeinschaft for financial support.

Authors' adresses:

Irene Luise Gügel, Ludwig-Maximilians-Universität München, Fakultät für Biologie, Department I, Biodiversität, / Anthropologie, Richard-Wagner-Str. 10, 80333 München, FRG

Karl-Heinz Kunzelmann, Poliklinik für Zahnerhaltung und Parodontologie, Goethestr. 70, 80336 München, FRG

Correspondence to:

Irene Luise Gügel, phone: +49-89-21806706, fax: +49-89-21806719; iguegel@lrz.uni-muenchen.de

BIBLIOGRAPHY

AINAMO, J., 1972, Relationship between occlusal wear of the teeth and periodontal health. *Scandinavian Journal of Dental Research* 80, p. 505.

BERRY, D. C. & POOLE, D. F. G., 1976, Attrition: possible mechanisms of compensation. *Journal of Oral Rehabilitation* 3, p. 201-206.

BEYNON, A. D., 1987, Replication technique for studying microstructure in fossil enamel. *Scanning Microscopy* 1, p. 663-669.

DE GEE, A. J., PALLAV, P., DAVIDSON, C. L., 1986, Effect of abrasion medium on wear of stress-bearing composites and amalgam in vitro. *Journal of Dental Research* 65, p. 654-658.

GÜGEL, I. L., GRUPE, G., KUNZELMANN, K.-H., 2001, Simulation of dental microwear: Characteristic traces by opal phytoliths give clues to ancient human dietary behavior. *American Journal of Physical Anthropology* 114, p. 124-138.

KUNZELMANN, K.-H., 1997, Verschleißanalyse und -quantifizierung von Füllungsmaterialien in vivo und in vitro. München: Habilitationsschrift LMU.

LARSEN, C. S., 1997, Masticatory and non-masticatory functions. In *Bioarchaeology: interpreting behavior from the human skeleton*, edited by C. S. Larsen. Cambridge University Press, p. 227–269.

LOMBARDI, A. V., 1982, The adaptive value of dental crowding: A consideration of the biologic basis of malocclusion. *American Journal of Orthodontics* 81, p. 8-42.

MEHL, A., GLOGER, W., KUNZELMANN, K.H., HICKEL, R., 1997, A new optical 3-D device for the detection of wear. *Journal of Dental Research* 76, p. 1799-1807.

MOLLESON, T., JONES, K., JONES, S., 1993, Dietary change and the effects of food preparation on microwear patterns in the Late Neolithic of abu Hureyra, northern Syria. *Journal of Human Evolution* 24, p.455-468.

MOLNAR, S., 1971, Human tooth wear, tooth function and cultural variability. *American Journal of Physical Anthropology* 34, p.175-190.

MOLNAR, S., 1972, Tooth wear and culture. *Current Anthropology* 13, p. 511-526.

MOLNAR, S., MCKEE, J. K., MOLNAR, I. M., PRZYBECK, T. R., 1983, Tooth wear rates among contemporary australian aborigines. *Journal of Dental Research* 62, p. 562-565.

POWELL, M. L., 1985, The Analysis of dental wear and caries for dietary reconstruction. In *The Analysis of prehistoric diets,* edited by R. J. jr. Gilbert, J. H., Mielke. Academic Press, Inc. p. 307-338.

ROSE, J. C., CONDON, K. W., GOODMAN, A. H., 1985, Diet and Dentition: Developmental Disturbances. In *The Analysis of prehistoric diets*, edited by R. J. jr. Gilbert, J. H., Mielke. Academic Press, Inc. p. 281-306.

ROSE, J. C. & UNGAR, P. S., 1998, Gross dental wear and dental microwear in historical perspective. In *Dental Anthropology, Fundamentals, Limits and Prospects*, edited by K. W. Alt, F. W. Rösing, M. Teschler-Nicola, Springer, p. 349 - 386.

SMITH, B. H., 1984, Patterns of molar wear in hunter-gatherers and agriculturalists. *American Journal of Physical Anthropology* 63, p. 39-56.

TEAFORD, M.F., LYTLE, J.D., 1996, Brief communication: Diet-induced changes in rates of human tooth microwear: a case study involving stone-ground maize. *American Journal of Physical Anthropology* 100, p. 143-147.

VÖTTER, W., 1973, Zeitliche und Geographische Unterschiede in der Abrasion der Zähne. Dissertation, Ludwig-Maximilians-Universität München.

WOLPOFF, M. H., 1971, Interstitial wear. *American Journal of Physical Anthropology* 34, p. 205-228.

BRIDGING THE GAP BETWEEN ACHEOLOGICAL DATASETS AND DIGITAL REPRESENTATIONS

Hervé DELINGETTE

Résumé: Dans cet article, sont décrites les méthodes mais aussi les difficultés pour créer une représentation numérique des données archéologiques. Dans un premier temps, nous exposons les principaux avantages de l'utilisation d'une base de données archéologiques sous forme entièrement numérique : la dissémination dans le temps et l'espace, la manipulation de grande quantité d'informations et l'accès à des mesures objectives. Dans un deuxième temps, nous présentons les techniques existantes pour la création de modèles géométrique d'objets ou sites archéologiques. On distingue ainsi la phase d'acquisition, de modélisation et d'édition de ces représentations numériques.

Abstract: In this article, we describe the methods but also the technical barriers for the creation of digital representation of archeological datasets. First, we give the main incentives for the exploitation of an archeological database that is entirely stored on a digital format : communication through time and space, handling of large datasets and computation of objective and reproductible measurements. Then, we detail the existing methods for the creation of geometric models of archeological artifacts or sites. We distinguish between the acquisition, modeling and editing of these digital representations.

INTRODUCTION

1 Digital Representations

In this paper, we present a general framework for building digital representations in the context of archeological excavations. The word « digital representation » is taken in a very broad sense and can be defined as any information that can be stored in a digital manner (with bits and bytes), i.e. inside a computer.

More precisely, the type of information can be any of the following :

Geometric information that describes the shape of an object (artefacts, surface layer,...). A common way of storing this information consists in describing a mesh, ie a list of 3D points and a list of connexion (edges) between these points.

Appearance information that describes how the objects looks like, either in its current or original format. This information can stored as a picture (taken by a digital camera) or a texture. A texture is an appearance information that is related to a geometric information.

Mechanical information that describes if objects are soft or hard. To quantify the stiffness of a material, one can provide the Young Modulus (E) as a physical value which is universally understood.

Position Information that indicates how an object is positionned in space with respect to a given coordinate system or relatively to another object

Semantic information that provides any other information (such as age, origins,...)

These information can be stored either in a qualitative or quantitative manner. For instance, a qualitative way of describing the geometry of an object is « its is blobby structure slighly flat on its top part » while a quantitative way consists in describing 3D triangulated mesh with the 3D coordinates of each point.

1.1 Motivations

The use of digital representations is of great interest in the field of archeology and more precisely for the management of archeological sites and objects. In fact, there has been many examples in the past few years of the creation and management of digital contents in this field, most of them motivated by the following arguments :

Preservation issue: A digital representation can be duplicated and archived instantaneously. By creating numerous digital copies of given artefacts and by archiving them all other the world, we can increase by a substancial amount the likelihood that its existence will be transmitted to future generations.

Ubiquity issue: with the development of world-wide computer networks (such as Internet), it has become trivial to share digital information accross the planet. This can be obviously beneficial for the scientific community regarding the exchange of artefacts, but also for the transmission of knowledge from the community to the general public through the existence of virtual museums.

Objective measurements issue: by using quantitative description of geometry and texture, it is possible to produce objective and reproductible measurements (such as computing volumes, distances and angles). It also helps to create gold standards for methodologies that can be largely accepted and used in the community.

Information management issue: considering on one hand the huge amount of information created by a cave excavation and on the other hand the complexity of this information (see 1.1), it becomes obvious that computer databases are the only sustainable way to exploit this pile of information. By coupling quantitative measurements (shape and position) with semantics, new findings can be foreseen with appropriate data mining software.

1.2 Limitations

Most arguments in favor of switching from physical to digital representation rely on a single assumption : that a physical object can be described by a set of sentences (semantics) and figures. This is basically not true. A digital representation is and will always be a specific and restrictive interpretation of a real object. No matter how many words or figures we use, the real object encapsulates much more information that can be stored within a single computer. This single fact could prevent the use of digital representations for any scientific activity.

However, the tools by which we currently analyse archeological artefacts are already limited in nature : our eyes, hands and tools can be of poor accuracy depending on the task to perform. Therefore, the real issue behind the creation of digital representations is to create « sufficiently accurate » copies of a real object. The amount of accuracy is typically driven by the application pursued. For instance, for comparing the shapes of two artefacts or to compute the volume of an endocranium, it is certainly not necessary to reach sub-millimeter accuracy. For the sake of preservation however, a very high resolution and accuracy should be reached in order to capture the most subtle details that could later on, reveal an important finding.

2 Creating Digital representations from real objects

In this section, the issue of creating digital representation is tackled with the objective to provide a general understanding of its difficulty. This is not a complete survey of existing techniques but rather a broad and schematic view of where we are now.

The idea of creating a digital replica of a real object has best described by Marc Levoy from the computer graphics department of Stanford University, who proposed the notion of « 3D fax machine »[3]. With this concept, one could place a real object with a box, push a button and then send its digital representation though a computer network to another machine that could rebuild the object identically. Far from being a science-fiction concept, a demonstration of this concept was first performed in February 1996.

In fact, the technology for the creating digital content has developed tremendously in the 90's with the occurrence of digital networks. However, it cannot be considered as mature since it has not entered yet an real industrial stage except for low-quality representations that are suitable for the entertainment industry.

We can summarized the creation of a digital representation from a real object by a three stage process (see Figure 1). First, it is important to note that the representation is itself an

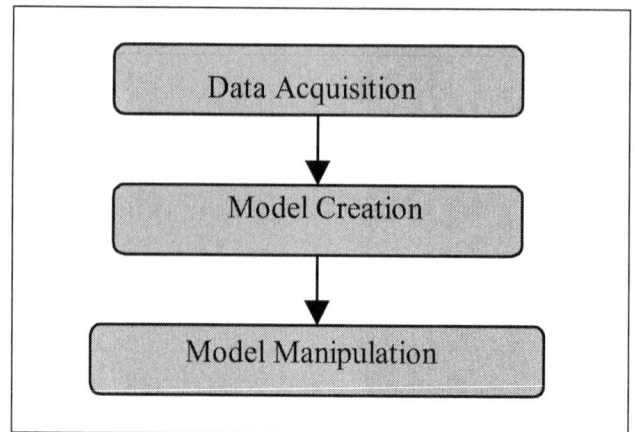

Figure 1 - Three stages for creation a digital representation of a real object

interpretation, a model of the real object. To create this model, it is often required to acquire some raw, to transform the data into a model and then to edit the model.

In the remainder, we mainly focus the description of these three stages for the creation of a geometric model of existing objects.

ACQUISITION OF SHAPE INFORMATION

The techniques for acquiring the shape of objects are commonly decomposed into passive and active techniques

Passive techniques, by definition, are based on physical parameters observed from the object without modifying its content and its nature. The most common approaches rely on digital cameras observing the object, ie counting the number of photons that it sends. When using two or more cameras, it is possible to mimic the human vision system by using the principle of stereo-vision : the same physical point appears in a different place in the right and left cameras and its disparity (difference of position) depends on the distance of that point from the image plane.

Other techniques borrowed from the computer vision community allows to reconstruct an object from a set of video-images (shape from motion) or from a set of focused images (shape from focus). The measurements that can be recovered from these methods may not be of high precision if the cameras are not calibrated. But the field of 3D photogrammetry and reconstruction from aerial images have shown excellent results for objects of large size. In figure 2, we see an example of 3D reconstruction of a large archeological site by the help of several cameras located around the object of interest (left). The different views are merged with the help of an operator and a cloud of points is created (right).

The second set of techniques are called active because they perturbate the object they are sensing. The mostly used active sensing method is called « laser range finding » . Laser range finder consists of a laser source and a camera and use a

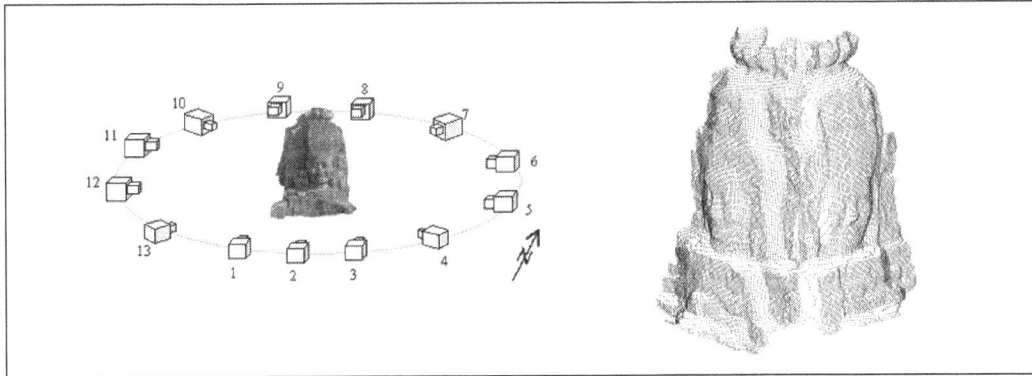

Figure 2 - Reconstruction of an archeological site (Ankor) performed with a photogrammetric technique at the polytechnical university of Zurich Pr. Grun)

Figure 3 - Acquisition of a statue using a specifically designed laser range finger (courtesy of Cyberware Corp. And Stanford University)

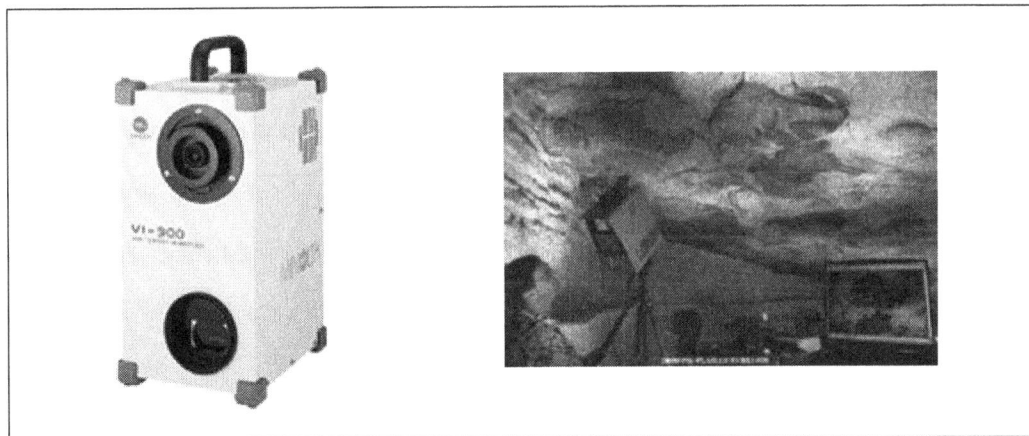

Figure 4 - Range scanning of a cave ceiling (courtesy of Minolta Corp.)

triangulation technique to measure the shape of a surface. There are several commercial products that are currently available since they often are fast, rather accurate and can be used by non-specialists. For many years, these range finders could only be used for small objects, but recent developments have allowed to scan large very tall objects like Michelangelo statue (see figure 3) or large parts of cave ceiling (see figure 4).

These two methods of active digital range scanning produce a dense map of 3D points in a few seconds (some of them even several times per second).

However these points cannot be used directly as a geometric model because there may be several undesirable artifacts :

Noise: the 3D points may be noisy (lack of accuracy)

Outliers: some 3D points may be completely wrong due to a failure of the triangulation algorithm

Missing data: because of the triangulation algorithm some parts of the object may be seen by the camera but not reached by the laser beam thus creating holes in the dataset.

Furthermore, as for passive techniques, a given acquisition allows only to reconstruct one side of an object. Therefore this is a need to stitch different sets of 3D points together in order to capture the whole object shape.

Finally, another active acquisition method can be used for small objects (typically archeological artifacts) : medical imaging. More precisely, X-ray computed tomography also known as CT-scanner, allows to acquire in 3D the volume (and not only the surface) of an object. Skulls, for instance, have been CT-scanned for several years with the purpose of performing morphometric measurements (Subsol et al, 2000, Odin et al, this volume].

CREATING AND EDITING GEOMETRIC MODELS

The sets of 3D points produced by one of the methods described above cannot be used directly as a geometric model. Indeed, the stage of model creation has several purposes:

Data fusion: this is useful to build a complete representation of an object from several side views. This fusion requires to register each data to a common coordinate system. This task can be done automatically if the overlapping area between two adjacent views is large enough.

Filtering: as point-out before, there maybe be holes or outliers in these datasets.

Mesh fitting: a set of 3D points is often not sufficient to properly describe a shape. A mesh that includes the description of adjacency between points, must be created in most cases. The mesh can be described as a set of triangles or NURBS surface patches.

Compression: depending on the final task, it is often necessary to discards points for creating compact geometric representation for the purpose of visualisation.

It is important to remember that an object should not be represented by a single model but by a set of geometric models of various resolutions (multi-resolution or progressive meshes).

If the model creation stage consists in reformatting the data provided by the acquisition sensors, the edition stage aims at adding semantic information on these models. For instance, an expert user can interactively attach labels to parts of a surface model or draw salient curves or other landmarks. These additional information can later on be used to perform morphometric measurements or to compare the shapes of different objects.

CONCLUSION

Considering the rapid development of technologies for the creation of digital representations, the management of archeological datasets is likely to change in the near future. By allowing the creation of multiple copies of a, once before, unique invaluable object, the arrival of the digital age should have a very positive impact on the archeological community. One could even foresee the existence of a worldwide archeological database as a major milestone towards the better understanding of human heritage.

Address of the author:

Hervé Delingette, INRIA Sophia-Antipolis

2004 route des Lucioles,

06902 Sophia-Antipolis, France

Herve.Delingette@inria.fr

BIBLIOGRAPHY

Curless,B.,& Levoy,M., 1996, A Volumetric Method for Building Complex Models from Range Images , Computer Graphics (ACM SIGGRAPH 96 Proceedings), Los Angeles. http://graphics.stanford.edu/projects/faxing/

Odin, G., Quatrehomme, G., Subsol, G., Delingette, H., Mafart,B., & de Lumley M.A., 2001, Comparison of a Three-Dimensional and a Computerized Assisted Method for Cranio-Facial Reconstruction: Application to Tautavel Man. In *Proceedings of the XIV International Congress of Prehistoric and Protohistoric Science* , Liège (Belgium), this volume.

Subsol, G., Mafart,B., Méline,D., Silvestre,A. & de Lumley, M.A., 2000, Traitement d'images scanographiques appliqué à l'étude tridimensionnelle de l'évolution de la forme du crâne humain. In *L'identité humaine en question,* edited by P. Andrieux, D. Hadjouis, and A. Dambricourt-Malassé, editors, Collection Paléoanthropologie et Paléopathologie osseuse, Artcom press, Paris, pages 92-101.

VOLUME COMPUTATION OF ARCHAEOLOGICAL VESSELS

Robert SABLATNIG, Srdan TOSOVIC & Martin KAMPEL

Résumé: Un grand nombre de fragments de poterie sont découverts lors des fouilles archéologiques. Cependant, le décompte analytique et l'analyse de ces fragments sont un investissement en temps et en personnel considérable. Pour ces raisons, les auteurs proposent au point un système de documentation assisté par ordinateur dans lequel les fragments archéologiques forment la base d'une classification et d'une reconstruction semi-automatique. Le principal objectif de ce projet est de mettre au point une méthode de classification automatisée, basée sur la section de profil de l'objet orienté, soit la section transversale du fragment orienté par l'axe de rotation de la symétrie. Pour compléter le profil, une représentation tridimensionnelle de la poterie étudiée s'avère donc nécessaire. L'objectif final est d'obtenir un outil qui puisse aider les archéologues dans leur processus d'archivage. Les auteurs décrivent l'état d'avancée de leurs recherche pour la mise au point d'un processus automatisé d'archivage et d'acquisition 3D qui réponde aux besoins des archéologues.

Abstract: At excavations a large number of sherds of archaeological pottery is found. Since the documentation and administration of these fragments represent a temporal and personnel effort, we construct a computer aided documentation system for archaeological fragments to form the basis for a subsequent semi-automatic classification and reconstruction. The main technical goal of this project is to perform an automated classification based on the profile section of the oriented object, which is the cross-section of the fragment in the direction of the rotational axis of symmetry. To achieve the profile, a 3d-representation of the object is necessary. The final aim is to provide a tool that helps archaeologists in their archivation process. This paper gives an overview about an automated archivation process and 3d-acquisition with respect to archaeological requirements.

INTRODUCTION

New technologies are introduced to old research areas and provide new insights for both researchers and people interested in this field. This statement can be proved especially in the field of archaeology, since there are many researchers in that area who already use new technologies and there are many people interested in the field of archaeology since so-called archaeo-parks have an increasing number of visitors (Fowler et al,1993, Woodruff et al, 2001). Motivated by the requirements of the present archaeology, we are developing an automated system for archaeological classification of ceramics. Ceramics are among of the most widespread archaeological finds, having a short period of use. A large number of ceramic fragments are found at nearly every excavation and have to be photographed, measured, drawn (Figure 1) and classified.

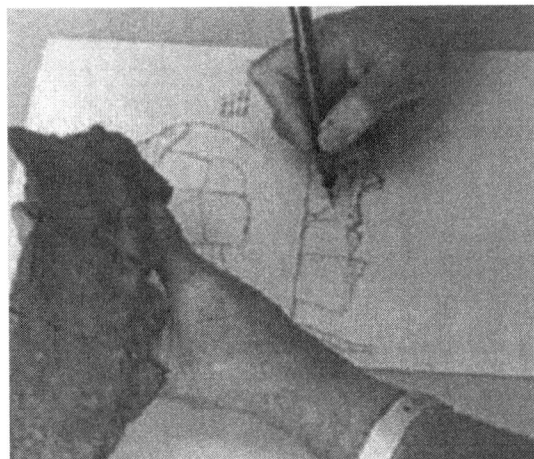

Figure 1 - Manual Drawing of fragments

Because the conventional methods for documentation and classification are often unsatisfactory (Orton et al, 1993), we are developing an automated archivation system (Kampel et al, 1999) that tries to combine traditional classification methods with new techniques in order to get an objective classification scheme.

Late-Roman burnished ware, which was found during the excavations from 1968 to 1977 in the legionary fortress of Carnuntum in Austria (Kandler, 1975 & 1979), was chosen as the basis for our research (Grünewald et al, 1979 & 1986). In addition to these sherds we enlarged our material basis with published pieces from other pannonian sites.

The purpose of classification is to get a systematic view and order on the excavation finds: treating every sherd as unique inhibits a clear view of the material (like not seeing the wood for the trees). Archaeological classification is traditionally done by typology: more or less defined forms are identified to possess certain significance and then addressed as "types". These "types" can be used as a sort of "label", which simplifies comparative the scientific field (Rice, 1987, Woodruff, 1991, Sinopoli, 1991, Bernbeck, 1997, Eggert et al, 2001). Furthermore, with the recognition of vessel types, patterns can be recognized. Hence, classification provides the basis for statistical analysis. But archaeologists often leave their typologies at an "impressionistic" or indefinite level, because their main task is only to present new material. There have been many attempts to objectify and standardize shape description and classification - also in connection with systems for automated recording (Kampfmeyer, 1987, Steckner, 1989, Poblome et al, 1997) -, but in practical archaeological research most of the consequent formal and mathematical classification schemes did not find a wider reception or application because they are often too vague, abstract, reductionistic or unpracticable (Orton et al, 1993).

The attributes of a successful classification have been summarized by Orton and others (1993):

- objects belonging to the same type should be similar (internal cohesion)

- objects belonging to different types should be dissimilar (external isolation)

- the types should be defined with sufficient precision to allow others to duplicate

- it should be possible to decide which type a new object belongs to

In order to achieve these aims our classification scheme of the vessel form is based on:

- absolute measurements and ratios
- segmentation of the profile line

The first steps are the measurements of the following parameters: rim diameter, bottom diameter, height, x- and y-values in all segmentation points. With these measurements a variety of ratios can be calculated. A special choice of these ratios is in each case characteristic for one vessel type; for example the ratio rim diameter to height or maximum diameter of the neck to maximum diameter of the belly.

The second aspect of classification is the segmentation of the vessel into its parts, the so-called primitives. The basis for this segmentation is the outer profile line that means the profile line along the outside of the vessel. The curve is described by means of a modified Cartesian system of co-ordinates. The x-axis of the system of co-ordinates lies in the orifice plane; the y-axis corresponds with the axis of rotation. The position of the curvature points is defined by means of x- and y-values. The profile line is situated in the right lower quadrant, so that not only complete vessels, but also rim fragments can be described (Kampfmeyer, 1987).

The profile of the vessel is composed of several segments, the so called primitives, for example: rim, neck, shoulder etc. If there is a corner point, that is a point, where the direction

of the curve changes "substantially", the segmentation point is obvious. If there is no corner point, the segmentation point has to be determined mathematically (Shepard, 1956, Woodruff et al, 1991).

Several points characterize the curve. Figure 2 shows the segmentation scheme of an S-shaped vessel as an example:

- **IP** *inflexion point*: point, where the curvature changes its sign, that means where the curve changes from a left turn to a right turn or vice versa;

- **MA** *local maximum*: point of vertical tangency; point, where the x-value is bigger than in the surrounding area of the curve;

- **MI** *local minimum*: point of vertical tangency; point, where the x-value is smaller than in the surrounding area of the curve;

- **OP** *orifice point*: outermost point, where the profile line touches the orifice plane;

- **CP** *corner point*: point, where the curve changes its direction substantially;

- **BP** *base point*: outermost point, where the profile line touches the base plane;

- **RP** *point of the axis of rotation*: point, where the profile line touches the axis of rotation;

- **SP** *starting point*: in case of vessels with a horizontal rim: innermost point, where the profile line touches the orifice plane;

- **EP** *end point*: in case of fragments: arbitrary point, where the profile line ends

Figure 3 - One piece-vessel (a) and Two-piece vessel (b)

On the basis of the number and characteristics of the segments three kinds of vessels can be identified:

1. One-piece vessels (Figure 3a): These vessels consist of only one main segment. Their sides extend continuously inward or outward without reaching a point of vertical tangency;

2. Two-piece vessels (Figure 3b): These vessels consist of two main segments: upper part and lower part;

3. Multi-piece vessels (Figure 2):These vessels consist of three or more main segments. A special kind of them are the so-called S-shaped vessels, which are composed of neck, shoulder and belly.

This paper is organized as follows: we started with a short introduction into archaeological classification and explained the criteria, which are used to establish a classification system. Section 2 deals with shape from silhouette for reconstruction of whole archaeological objects. In section 3 shape from structured light for the reconstruction

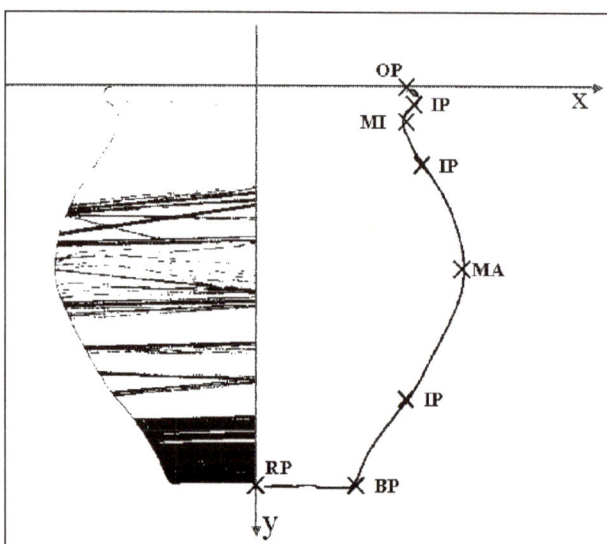

Figure 2 - S-shaped vessel: profile segmentation scheme

of fragments is presented. A laser ranges sensor is described in section 4. Results are given at the end of each section. This paper gives an overview about 3D acquisition methods used for the reconstruction of archaeological finds. We conclude with a summary and an outlook on future research.

Shape from Silhouette for 3D modeling of complete objects

Shape from Silhouette is a method of automatic construction of a 3D model of an object based on a sequence of images of the object taken from multiple views, in which the object's silhouette represents the only interesting feature of an image (Szeliski, 1993, Potsemil, 1987). The object's silhouette in each input image corresponds to a conic volume in the object real-world space (see Figure 4). A 3D model of the object can be built by intersecting the conic volumes from all views.

Shape from Silhouette is a computationally simple algorithm — it employs only basic matrix operations for all transformations — and it requires only a camera as equipment, so it can be used to obtain a quick initial model of an object, which can then be refined by other methods. It can be applied on objects of arbitrary shapes, including objects with certain concavities (like a handle of a cup), as long as the concavities are visible from at least one input view. It can also be used to estimate the volume of an object. The shape from Silhouette algorithm used is described in detail in (Tosovic, 2000).

The acquisition system (Kampel, 2000) consists of the following devices:

• a monochrome CCD-camera with a focal length of 16 mm and a resolution of 768x576 pixels

• a turntable with a diameter of 50 cm, whose desired position can be specified with an accuracy of 0.05 degrees

An important issue is the illumination of the object observed, which should be clearly distinguishable from the background, independent from the object's shape or the type of its surface. For that reason back-lighting (Haralick et al, 1991) is used. A large (approx. 50x40 cm) rectangular lamp is put behind the turntable (as seen from the camera). In addition, a white piece of paper, larger than the lamp, is put right in front of the lamp, in order to make the light more diffuse. The whole system is protected against the ambient light by a thick black curtain.

Prior to any acquisition, the system is calibrated in order to determine the inner and outer orientation of the camera and the rotational axis of the turntable. The calibration method used was exclusively developed for the Shape from Silhouette algorithm used and it is described in detail in (Tosoic, 1999) and (Kampel, 2000). Figure 5 shows the reconstructed 3D models of the two pots from three sides. For these models octree of resolution of 256^3 voxels was built, based on input images from 36 views.

The results with both synthetic and real input data show that there is a certain minimal octree resolution required to obtain an accurate model of an object, especially for highly detailed objects, like the two pots used for tests with real images. Concerning the number of input views used for obtaining a model of an object, it turned out that beginning from 12 views, the constructed model does not change significantly. In our

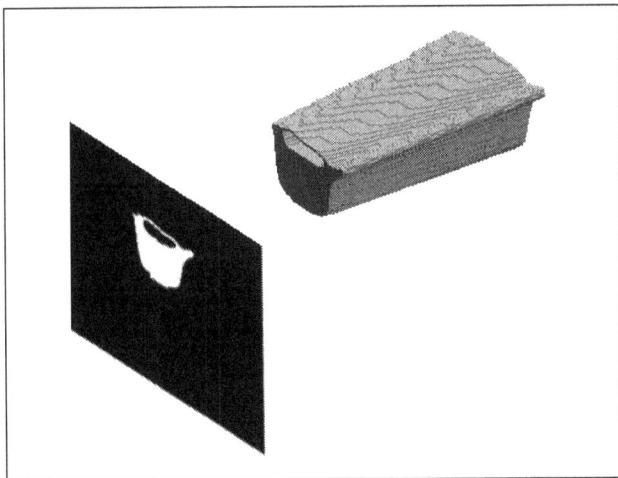

Figure 4 - Image silhouette and the corresponding conic volume

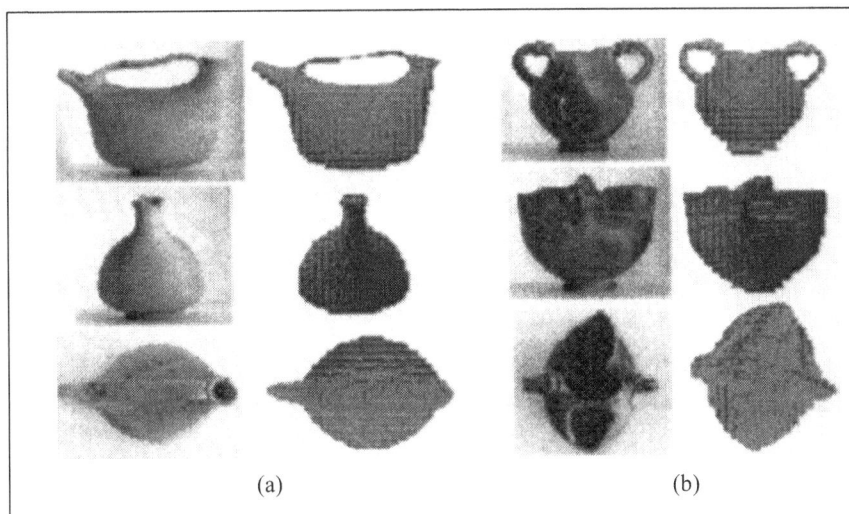

(a)   (b)

Figure 5 - Constructed models of real objects in different voxel resolution

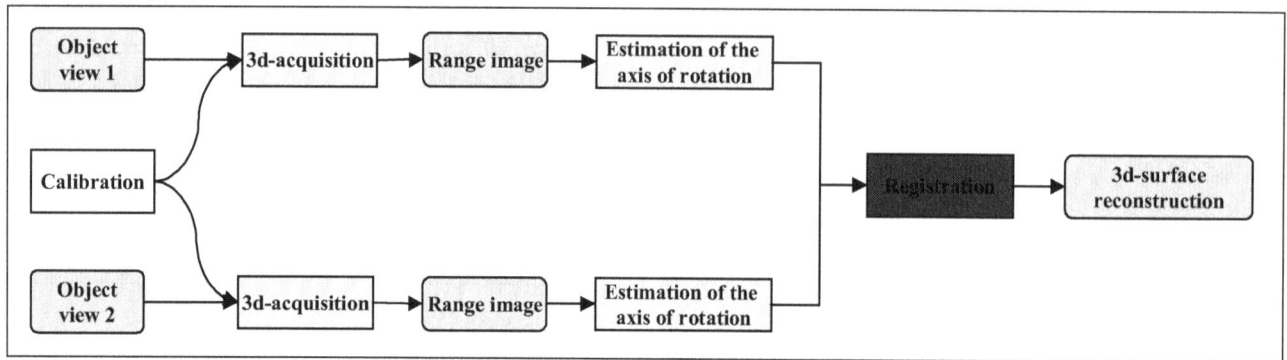

Figure 6 - 3d surface reconstruction overview

tests the octrees built from 12 views were almost the same as the ones built from 36 views, except that they took much less time to construct.

The results with synthetic data, where we had a perfect transformation matrix, showed that the error in the dimensions of the model lies within or is slightly higher than the error introduced through the minimal voxel size. The error with real data depends additionally on the accuracy of the calibration algorithm. The results also showed that the algorithm works much better with oval objects, i.e., with objects that do not have completely flat surfaces or sharp edges.

Coded light for 3D reconstruction of fragments

Our documentation system for archaeological fragments is based on the profile, which is the cross-section of the fragment in the direction of the rotational axis of symmetry. Hence the position of a fragment (orientation) on a vessel is important. To achieve the profile, a 3d-representation of the object is necessary.

Archaeological pottery is assumed to be rotationally symmetric since it was made on a rotation plate. With respect to this property the axis of rotation is calculated using a Hough inspired method (Ben-Yacoub et al, 1997). To perform the registration of the two surfaces of one fragment, we use a-priori information about fragments belonging to a complete vessel: both surfaces have the same axis of rotation since they belong to the same object.

Figure 6 gives an overview of a 3d-surface reconstruction from two object views. The first step consists of sensing the front- and backside of the object (in our case a rotationally symmetric fragment) using a calibrated 3d-acquisition system. We register the resulting range images by calculating the axis of rotation of each view and bringing the estimated axes into alignment. The method is described in detail in (Kampel, 1999).

In our acquisition system the stripe patterns are generated by a computer controlled transparent Liquid Crystal Display (LCD 640) projector. The light patterns allow the distinction of 2^n projection directions. Each direction can be described uniquely by a n-bit code. A CCD-camera is used for acquiring the images.

The projector projects stripe patterns onto the surface of the objects. In order to distinguish between stripes they are binary encoded. The camera grabs gray level images of the distorted light patterns at different times. With the help of the code and the known orientation parameters of the acquisition system, the 3d-information of the observed scene point can be computed. This is done by using the triangulation principle. The image obtained is a 2D array of depth values and is called a range image (Figure 7).

To find out if the method is working on real data we used a totally symmetric small flowerpot with known dimensions and took a fragment which covered approximately 25% of the original surface. The range images of the front- and back-view consisted of approximately 10.000 surface points each

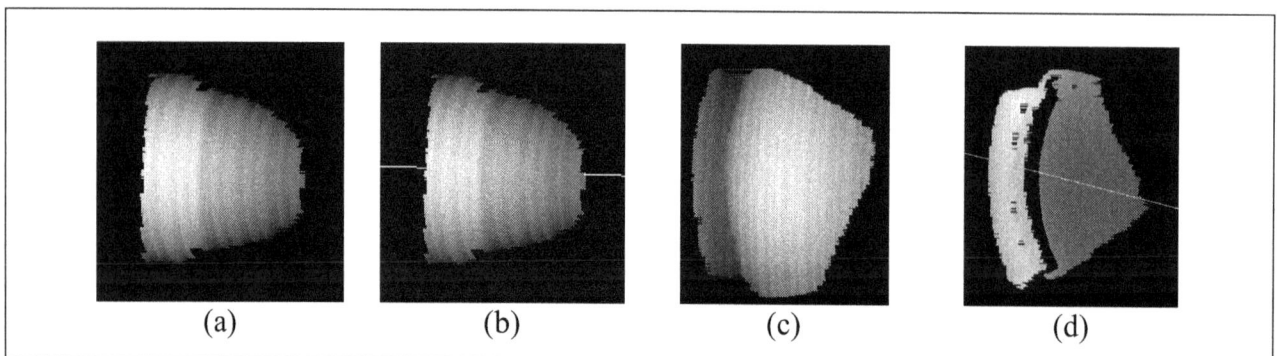

Figure 7 - Front- and back-view (range images) and their axis of rotation of a flowerpot (a, b) and an archaeological fragment (c, d).

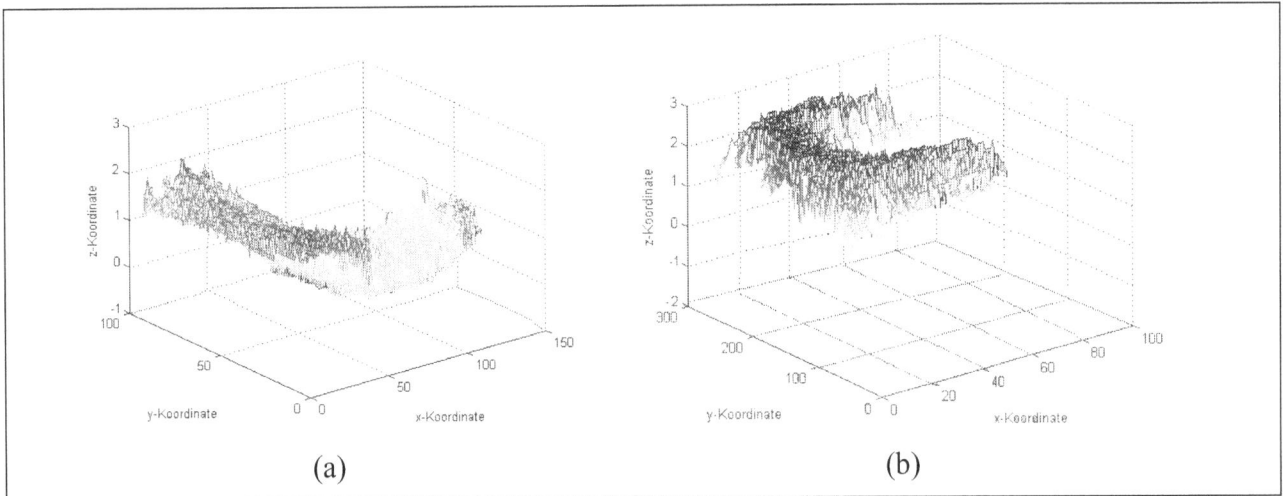

Figure 8 - Distribution of δ for registered flowerpot (a) and archaeological fragment (b)

(Figure 7a,b). The mean distance d between the surfaces is 5.64mm and the registration error d=1.42mm. The distribution of the registration error delta for the flowerpot is shown in Figure 8a. The registration error increases towards the top of the pot, because of the irregularity of the distance between the surfaces at that region since the flowerpot has an edge (upper border) where inner and outer surface are not parallel.

Figure 7c and d show the front-view, back-view and the axis of rotation of a real archaeological fragment. Registration tests with this fragment resulted in registration errors of approximately d=1.7mm and a mean distance of d=5.8mm.

Figure 8b shows the distribution of d of a registered archaeological fragment. Marginal peaks are caused by shadow regions of the back-view (see Figure 7d) at the border of the fragment, where either no range data is processed or the range information is unreliable. The increase of the registration error d reflects the uneven surface of the fragment.

Further problems that arise with real data are symmetry constraints, i.e. if the surface of the fragment is too flat or too small; the computation of the rotational axis is ambiguous (worst case: sphere) which results in sparse clusters in the Hough-space, which indicate that the rotational axis is not determinable.

Laser range sensor

The acquisition method used for estimating the 3d-shape of objects is shape from structured light, based on active triangulation (fig 9). The camera is positioned between the two lasers facing the measurement area. The complete system consists of:

- 1 turntable with a diameter of 50 cm, which can be rotated about the z-axis, used to move the object of interest through the acquisition area.

- 2 red lasers to illuminate the scene, one mounted on the top (distance to rotation plane is 45 cm), one beside the turntable (distance to the rotation center is 48 cm). Both lasers are extended with cylindrical lenses to spread the laser beam into one illuminating plane. The laser light plane intersects with the object surface, forming one laser stripe.

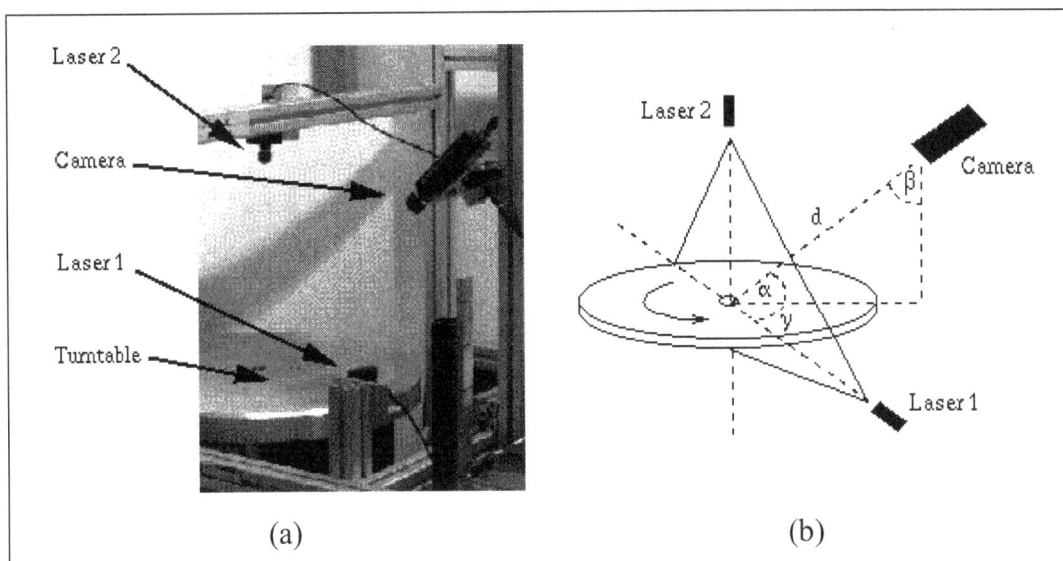

Figure 9 - Acquisition system with the complete hardware setup (a) and its geometric arrangement (b).

- 1 CCD-camera (b/w) with a 16 mm focal length, a resolution of 768x572 pixels, and a distance of 40 cm to the rotation center. The angle between the camera normal vector and the rotation plane is approx. 45 degrees. A frame grabber card is used to connect the camera to a PC.
- 1 Intel Pentium PC under Linux operating system.

An iterative process for 3d surface reconstruction in static environments is defined by the following steps, which depicts this process:

- Image acquisition: The scene is captured by the CCD-camera. The result is a greyscale- image, which shows the intersection between the laser plane and the object that is a line.
- Feature extraction: The line shown in the camera image is extracted. The result is a set of 2d points.
- Registration: The set of 2d points extracted in the previous step is transformed from the world coordinate system in its object coordinates.
- Integration: Each registered point is integrated into the existing model computed and integrated at the previous iterations of the acquisition process.
- Next View Planning: The next viewing angle is computed based on the algorithm shown in the previous section and the turntable moves to the calculated absolute angle. The process repeats until the turntable revolves one complete rotation.
- 3d-model visualization of the reconstructed surface.

Figure 10 shows the reconstruction of the head of an amphora after the acquisition process. The chosen angles are between 4 and 12 degrees. The analysis of the reconstruction data shows that the axis of symmetry is switched from the center of rotation by 1.8mm in x-direction and 2.1mm in y-direction. In order to scan the whole image 36 steps were necessary. Figure 10b shows the rendered object after the surface reconstruction

process. The visualization results from a modified z-buffer-algorithm.

CONCLUSION AND OUTLOOK

We have proposed a prototype system for 3d acquisition of archaeological fragments. The work was performed in the framework of the documentation of ceramic fragments. The methods proposed have been tested on synthetic and real data with reasonably good results. It is part of continuing research to improve the results from multiple, various objects since the technique has some drawbacks. The first one refers more to the calibration algorithm, which makes many simplifying assumptions about the acquisition system.

The achieved fragment representations, the first part of an automated system for classification of archaeological fragments, are the input of the second part of the system, classification. The classification will be solved in the high dimensional real space and therefore the uniqueness and the high precision of the profile representation are very important. The current work focuses on finding a unique orientation of profiles and on the final identification of vessels.

For archaeological applications, the object surface has to be smoothed in order to be applicable to texture mapping and therefore ceramic documentation, for classification, however, the accuracy of the method presented is sufficient since the projection of the decoration can be calculated and the volume estimation is much more precise than the estimated volume performed by archaeologists.

Acknowledgements

This work was partly supported by the Austrian Science Foundation (FWF) under grant P13385-INF, the EU under grant IST-1999-20273 and the Austrian Federal Ministry of Education, Science and Culture

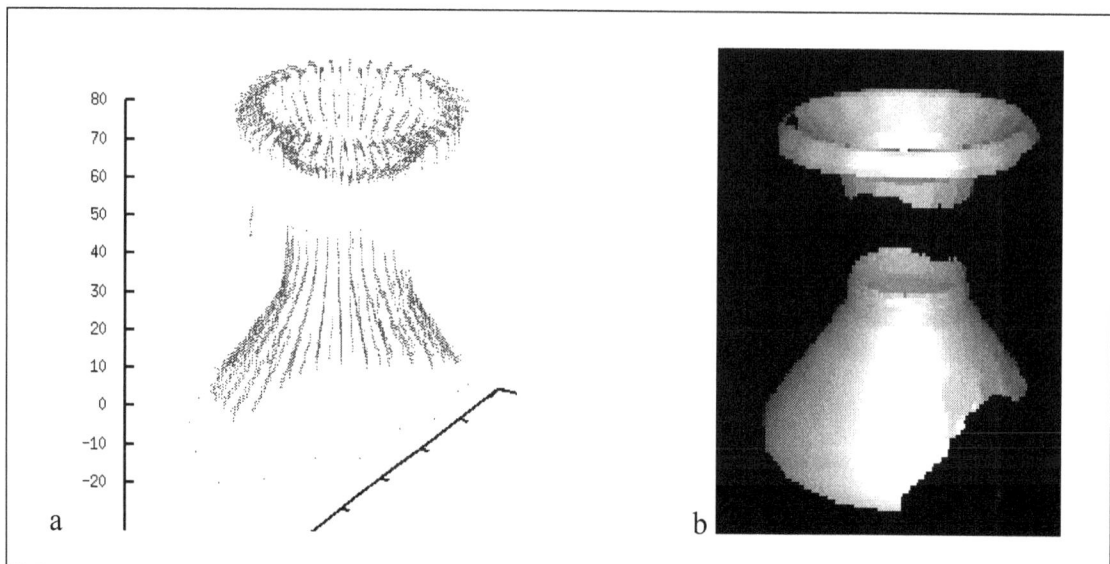

Figure 10 - Reconstruction of pottery

Author's addresses:

Pattern Recognition and Image Processing Group

Institute for Automation, Vienna University of Technology,

Favoritenstr. 9/183/2, A-1040 Vienna, Austria

Fax: +43 (1) 58801 183 92;

email: {sab, kampel}@prip.tuwien.ac.at

BIBLIOGRAPHY

ADAMS, W.Y., AND ADAMS, E.W.,1991, *Archaeological Typology and Practical Reality. A Dialectical Approach to Artifact Classification and Sorting.* Cambridge.

BEN-YACOUB,S.,&, MENARD,C.,1997, Robust Axis Determination for Rotational Symmetric Objects out of Range Data. IN BURGER. ,W, & BURGE, M., EDITORS, *21st Workshop of the OEAGM*, pp.197-202, Hallstatt, Austria.

BERNBECK, R., 1997, *Theorien in der Archäologie*, Tübingen and Basel.

EGGERT, M.K.H., 2001, *Prähistorische Archäologie. Konzepte und Methode*, Tübingen and Basel.

FOWLER, P., AND BONIFACE, P., 1993, Heritage and Tourism in *"The Global Village"*. London.

GRÜNEWALD, M., 1979, Die Gefäßkeramik des Legionslagers von Carnuntum (Grabungen 1968-1977). In *DerRömische Limes in Österreich* 29, pages 74–80.

GRÜNEWALD, M., 1986, Ausgrabungen im Legionslager von Carnuntum (Grabungen 1969-1977). Keramik und Kleinfunde 1976 - 1977. In *Der Römische Limes in Österreich* 34, pp 10–11.

HARALICK, R.M.,& SHAPIRO, L.G., 1991, Glossary of computer vision terms. *Pattern Recognition* 241 : 69-93.

KAMPEL, M. & SABLATNIG, R., 1999, On 3d Modelling of Archaeological Sherds, In *Proceedings of International Workshop on Synthetic-Natural Hybrid Coding and Three Dimensional Imaging*, pp. 95-98.

KAMPEL, M. & TOSOVIC S., 2000, "Turntable calibration for automatic 3D-reconstruction," in *Applications of 3D-Imaging and Graph-based Modelling, Proceedings of the 24th Workshop of the Austrian Association for Pattern Recognition (ÖAGM)*, pp. 25-31.

KAMPEL, M., 1999, *"Tiefendatenregistrierung von rotationssymmetrischen Objekten"*, Master Thesis, Vienna University of Technology, Institute of Automation.

KAMPFMEYER, U., 1987, Untersuchungen zur rechnergestützten Klassifikation der Form von Keramik. *Arbeiten zur Urgeschichte des Menschen* 11. Frankfurt am Main.

KANDLER,M., 1975, *Anzeiger der österreichischen Akademie der Wissenschaften* 111 : 27 – 40.

KANDLER,M., 1979, *Anzeiger der österreichischen Akademie der Wissenschaften* 115 : 335–351.

MAVER, J., & BAJCSY, R., 1993, "Occlusions as a Guide for Planning the Next View," IEEE *Transactions on Pattern Analysis and Machine Intelligence* 15, pp.417-432.

ORTON, C, TYERS, P., VINCE, A., 1993, *Pottery in Archaeology.*

POBLOME, J., VAN DEN BRANDT, J., MICHIELS, B., EVSEVER, DEGEEST, R., & WAELKENS, M, 1997, Manual Drawing versus Automated Recording of Ceramics. In M. Waelkens editor, Sagalassos IV, Leuven., *Acta Archaeologica Lovaniensia Monographiae* 9, 533–538,

POTSEMIL, M., 1990,Generating octree models of 3D objects from their silhouettes in a sequence of images. *Computer Vision, Graphics, and Image Proecessing* 40: pp 68-84.

RICE, P.M., 1987, *Pottery Analysis: A Sourcebook.*

SABLATNIG, R.,& MENARD,C., 1996, Computer based Acquisition of Archaeological Finds: The First Step towards Automatic Classification. In P. Moscati / S. Mariotti, editor, *Proceedings of the 3rd International Symposium on Computing and Archaeology*, Rome, volume 1, pp 429– 446.

SHEPARD,A. O., 1956, Shepard. *Ceramics for Archaeologists.* Washington (9th reprint 1976).

SINOPOLI, C.M., 1991,.*Approaches to Archaeological Ceramics.* New York.

STECKNER, C.,1989, Das SAMOS Projekt. *Archäologie in Deutschland*, 1 : 16–21.

SZELISKI, R.,1993, Rapid octee construction from image sequences. *CVGIP: Image Understanding*,58 :23-32.

TOSOVIC, S., 1999, "Lineare Hough-Transformation und Drehtellerkalibrierung," Institute of Computer Aided Automation, Pattern Recognition and Image Processing Group, Vienna University of Technology, Austria, *Tech. Rep. PRIP-TR-59.*

TOSOVIC, S., 2000, Shape from Silhouette", Technical Report, PRIP-TR-64, Pattern Recognition and Image Processing Group, Institute for Computer Aided Automation, Vienna University of Technology.

WOODRUFF, A, AOKI, P, HURST, A,, AND SZYMANSKI, M, 2001, Electronic Guidebooks and Visitor Attention. In BEARMANAND D. GARZOTT F, editors, *Proceedings of the International Conference on Cultural Heritage and Technologies in the Third Millennium*, Milan, volume1, pages 437–454, 2001.

RECONSTITUTION EN TROIS DIMENSIONS D'UN SOL D'HABITAT PREHISTORIQUE: EXEMPLE DE LA CAUNE DE L'ARAGO (TAUTAVEL, PYRENEES-ORIENTALES, FRANCE)

Henry de LUMLEY, Caroline BUTOUR, Anne-Marie MOIGNE, Véronique POIS, Rachel VAUDRON

Résumé: La Caune de l'Arago est un site du Pléistocène moyen qui a déjà fait l'objet de plus de trente cinq années de fouilles qui ont donné lieu à de nombreuses analyses. Les résultats de ces diverses études ont été saisis dans la Base de Données « Matériel Paléontologique et Préhistorique » ou scannérisés pour constituer une bibliothèque numérique. L'important développement de l'outil informatique permet d'entreprendre la reconstitution en trois dimensions d'un sol d'habitat préhistorique et d'aborder la simulation de l'évolution géodynamique des dépôts quaternaires de la Caune de l'Arago.

Abstract: The Caune de l'Arago is a site from the Middle Pleistocene which has been the subject of excavations for more than thirty-five years. Those excavations have given place to many analysis. The results of these various studies have been seized in the database « Prehistoric and Paleontologic Material » or have been scanned to constitute a numerical library. The important development of informatic tools enables to undertake the three-dimensional reconstruction of prehistoric ground of settlement and to enter upon the simulation of the geodynamic evolution of the quaternary filling of the Caune de l'Arago.

INTRODUCTION

L'important développement de l'outil informatique au cours des dernières décennies a bouleversé le monde de la Préhistoire avec l'apparition de concepts informatiques extrêmement puissants qui ont ouvert de nouvelles voies de recherche. Désormais, l'étude des hommes préhistoriques et de leur mode de vie bénéfice des avantages techniques et méthodologiques apportés par l'Informatique : rigueur, rapidité, analyse des hypothèses, stockage des données de fouilles et des collections, pérennité de l'information, classification et analyse d'objets archéologiques, sélectivité, application des méthodes mathématiques et statistiques, réseaux, modélisation des reconstitutions archéologiques, simulation de systèmes archéologiques.

Fouillée depuis 40 années, la Caune de l'Arago a livré un important matériel archéologique à l'origine de nombreuses études. Les résultats des diverses analyses ont été saisis dans la Base de Données « Matériel Paléontologique et Préhistorique » ou scannés pour constituer une bibliothèque numérique. En plus de son grand intérêt archéologique, la Caune de l'Arago est un gisement exceptionnel dont le remplissage a conservé l'enregistrement de l'altération géochimique et des déformations synsédimentaires et postdépositionnelles.

A partir de l'ensemble de ces données, la reconstitution en trois dimensions d'un sol d'habitat préhistorique peut être entreprise. Pour cette étude, il est nécessaire d'établir les démarches fondamentales permettant de procéder :

1. à une restitution, en plan et en trois dimensions, des divers sols d'habitat préhistorique superposés dans l'état actuel.

2. à une restitution, en plan et en trois dimensions, des divers sols d'habitat préhistorique tels qu'ils devaient se présenter à l'origine, après correction des déformations postdépositionnelles.

PRESENTATION DU SITE

Creusée dans les calcaires à faciès urgoniens, la Caune de l'Arago est située sur la commune de Tautavel (figure 1). La grotte a conservé une puissante séquence stratigraphique. Les sédiments s'y sont accumulés aussi bien en période froide qu'en période tempérée, en période sèche qu'en période humide, et ceci depuis le stade isotopique 17 jusqu'au stade isotopique 5.

Son remplissage, constitué de sédiments plus ou moins stériles intercalés de plus de 30 niveaux d'occupation humaine, s'échelonne de –690 000 à –100 000 ans, sur une épaisseur de l'ordre de 14 mètres (figure 2).

Postérieurement, une partie du remplissage sédimentaire de ce site du Pléistocène moyen a subi une importante évolution géochimique très probablement en relation avec un amas de guano de chauves-souris. Les fluides ont percolé au sein de cette accumulation de matière organique. Rendus ainsi

Figure 1- Caune de l'Arago (Tautavel, Pyrénées-Orientales) – Situation géographique (d'après Lumley H. de et al., 1981).

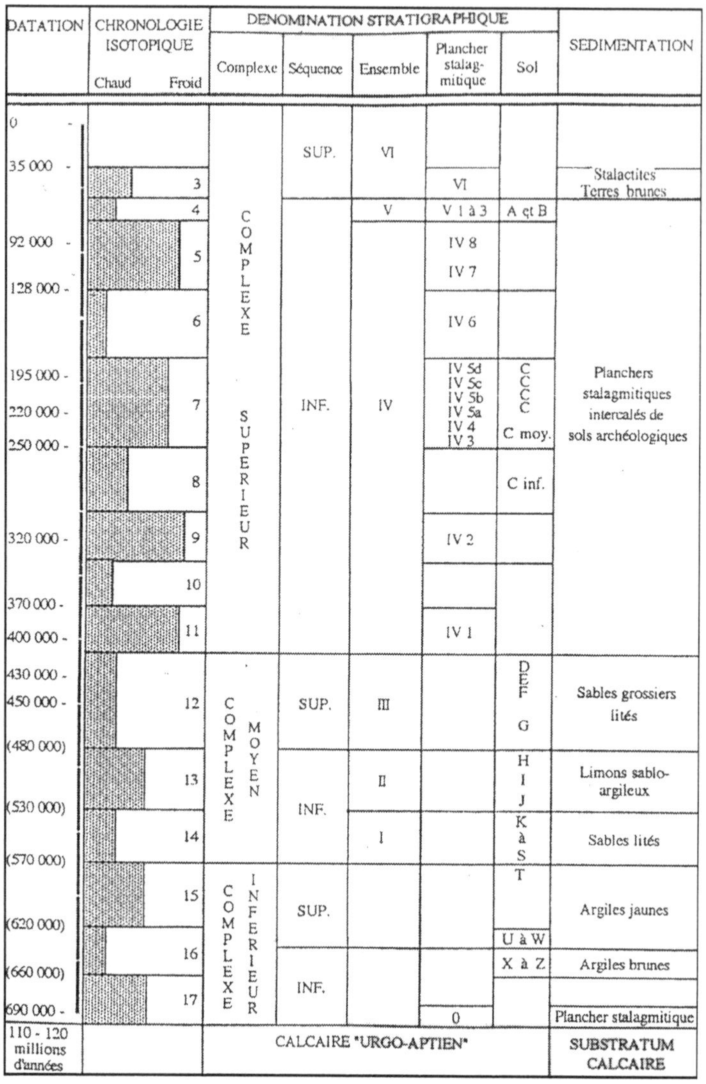

DATATION	CHRONOLOGIE ISOTOPIQUE (Chaud / Froid)	DENOMINATION STRATIGRAPHIQUE					SEDIMENTATION
		Complexe	Séquence	Ensemble	Plancher stalag-mitique	Sol	
0			SUP.	VI			
35 000	3				VI		Stalactites Terres brunes
	4	C O M P L E X E		V	V 1 à 3	A et B	
92 000	5				IV 8		
128 000					IV 7		
	6				IV 6		
195 000		S U P E R I E U R	INF.	IV	IV 5d / IV 5c / IV 5b / IV 5a / IV 4 / IV 3	C / C / C / C	Planchers stalagmitiques intercalés de sols archéologiques
220 000	7					C moy.	
250 000							
	8					C inf.	
320 000	9				IV 2		
370 000	10						
400 000	11				IV 1		
430 000		C O M P L E X E M O Y E N	SUP.	III		D E F	Sables grossiers lités
450 000	12					G	
(480 000)						H	
(530 000)	13		INF.	II		I J	Limons sablo-argileux
	14			I		K à S	Sables lités
(570 000)						T	
(620 000)	15	C O M P L E X E I N F E R I E U R	SUP.				Argiles jaunes
	16					U à W	
(660 000)			INF.			X à Z	Argiles brunes
690 000	17					0	Plancher stalagmitique
110 - 120 millions d'années		CALCAIRE "URGO-APTIEN"					SUBSTRATUM CALCAIRE

Figure 2 - Caune de l'Arago (Tautavel, Pyrénées-Orientales)
– Log stratigraphique synthétique du remplissage
(d'après Lumley H. de et al., 1981).

agressifs, ils ont conduit à la constitution d'une importante poche décarbonatée en forme d'entonnoir, par une succession de transformations minéralogiques et structurales distinctes. De ce fait, les niveaux archéologiques et les couches stratigraphiques de la Caune de l'Arago ne correspondent pas à leur conformation originelle qui était horizontale et située plus haut.

METHODOLOGIE

Restitution, en plan et en trois dimensions, des divers sols d'habitat préhistorique superposés dans l'état actuel

Les divers sols d'habitat préhistorique superposés dans l'état actuel sont individualisés à partir de la description des coupes stratigraphiques longitudinales et transversales, de l'étude de leurs recoupements et de l'exploitation des profils d'objets sur plans verticaux transversaux et longitudinaux. Cette étude permet de déterminer la liste du matériel archéologique contenu dans chaque sol.

Grâce à deux logiciels développés au sein du laboratoire, ARCNEW par J. Fruitet (1991) et ARCPROF par A. Canals (1993), il est possible de procéder à la réalisation de projections d'objets saisis dans la base de données, sur plans verticaux. Ces profils découpent le remplissage de la cavité karstique en tranches de 10 cm dans les sens longitudinal et transversal. Ils sont utilisés pour restituer les couches fouillées.

Le relevé sur le terrain des coupes stratigraphiques donne également une vision globale de la stratigraphie actuelle des dépôts fouillés. Visibles grâce à l'extraction des éléments archéologiques et des sédiments, les coupes correspondent à des profils sur lesquels sont aussi bien représentés le matériel archéologique en section que la sédimentologie avec les sables, argiles et limons, la bioturbation et la fracturation.

L'étude en parallèle de ces deux documents, projection d'objets (figure 3) et coupe stratigraphique (figure 4), permet, avant tout, de mettre en évidence une troncature du remplissage en relation avec l'ouverture de l'aven au plafond de la grotte et l'effondrement du porche original. La surface du remplissage n'apparaît, en effet, pas sur les projections

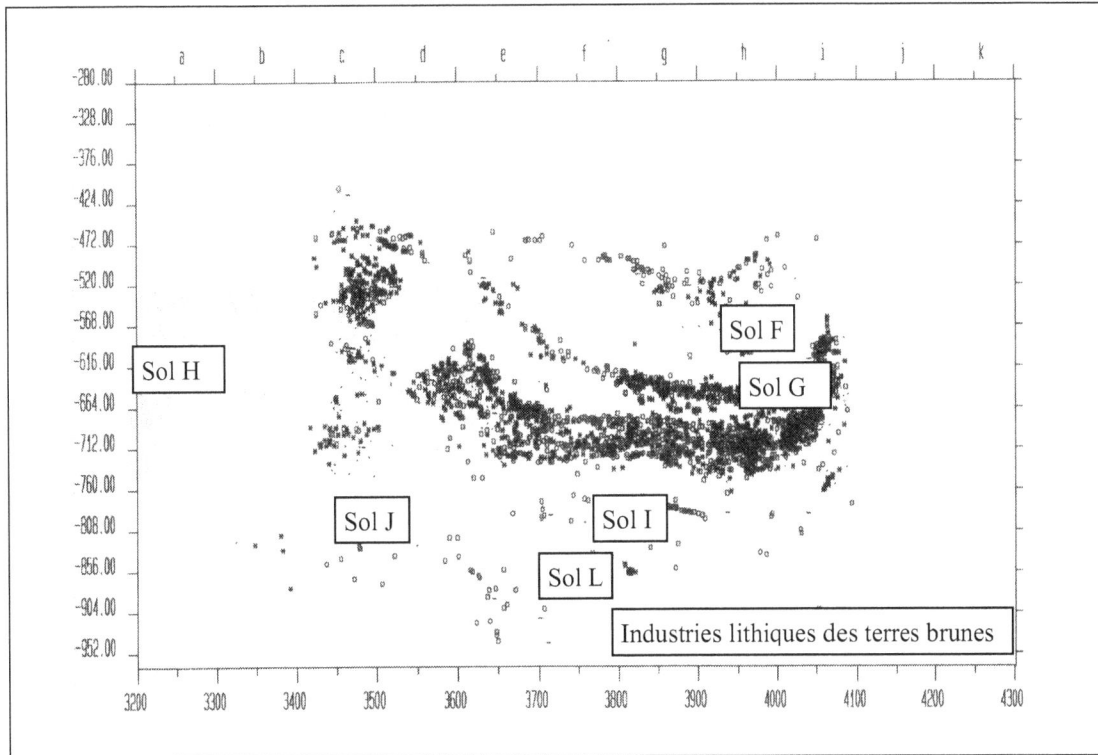

Figure 3 - Caune de l'Arago (Tautavel, Pyrénées-Orientales) –
Profil en bande 15 pour un Y compris entre 90 et 99 centimètres (Pois V., 1998).
*Légende- o industries lithiques, * ossements, - schistes, g galets, . pierres*

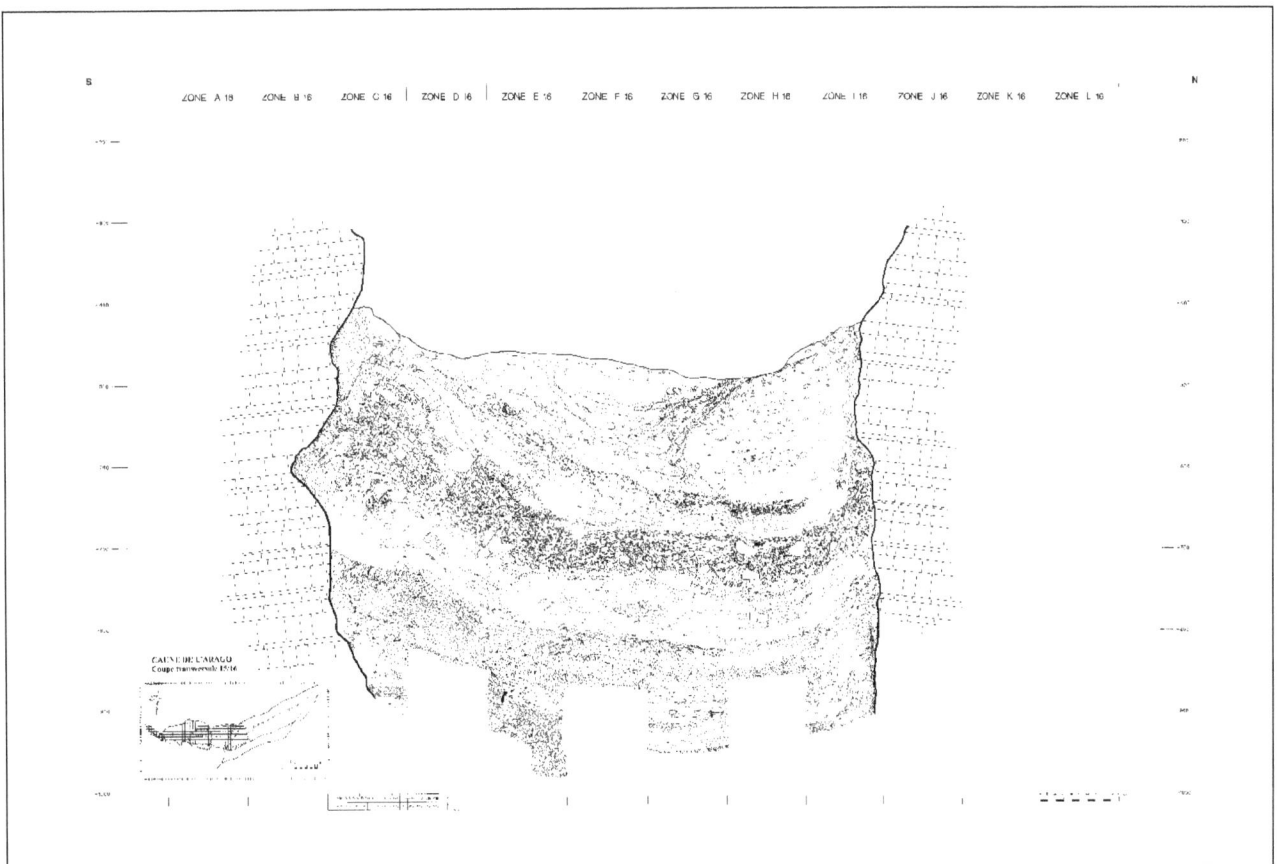

Figure 4 - Coupe stratigraphique transversale 15 / 16,
Caune de l'Arago (Tautavel, Pyrénées-Orientales) (d'après Lumley H. de et *al.*,1981).

d'objets, ni même le profil des parois. Pourtant, les phénomènes de parois nombreux à la Caune de l'Arago, expliquent la répartition du matériel archéologique à proximité.

Grâce aux coupes stratigraphiques, certains niveaux archéologiques distingués par la taille du matériel qu'ils contiennent sont mis en évidence par le relevé à l'échelle 1/10ième. La taille du matériel archéologique précisée sur les coupes stratigraphiques relevées à l'échelle 1/10ième est utilisée pour séparer des niveaux archéologiques. En effet, sur cette coupe, l'intersol F-G devient de plus en plus discret vers la paroi Est, à tel point qu'en bande I, le sol F repose directement sur le sol G. Or, les observations sur le terrain ont permis de constater que les objets de G étaient plus gros que ceux de F. La coupe stratigraphique met en évidence ce phénomène et permet de différencier F de G en raison de la présence d'objets avoisinant les 20 cm.

Il est, par ailleurs, intéressant de constater que les niveaux situés au-dessous du sol G se devinent plus qu'ils ne se voient en raison d'objets de plus en plus rares. Le relevé stratigraphique apparaît alors indispensable puisque c'est le litage des sables intercalés avec les limons qui permet de concrétiser la continuité stratigraphique entre des objets plus ou moins éloignés. Par ailleurs, il n'est pas rare d'observer des niveaux lenticulaires ou des lits de faible épaisseur, constitués soit de manganèse soit de phosphate, qui se distinguent dans le prolongement des lits sablo-limoneux et qui concrétisent de la même manière la continuité stratigraphique.

De la même façon, la nature du matériel archéologique, précisée cette fois-ci sur les projections d'objets, est également un bon indice de repérage des niveaux. Ainsi, le niveau archéologique H est représenté par la présence d'outils lithiques (o). Il est également caractérisé par une concentration de pierres (.) mêlées à des restes de faunes (*), en bande C. La partie supérieure du niveau archéologique I, par rapport à la partie inférieure, est non seulement plus riche en matériel archéologique mais également plus dense en industries lithiques qui constituent un horizon peu épais et continu au niveau de G. Le niveau J renferme des industries lithiques et de schistes (-) dans la poche de décarbonatation. Contre la paroi Ouest où l'altération n'est pas marquée, les pierres et les ossements sont nombreux. En (g), ce sont les galets. Le niveau K est peu épais et pauvre en matériel archéologique. Le niveau L se caractérise par une abondance de restes osseux (*) et quelques industries lithiques.

A l'aplomb des bandes E et G, du fait d'une fouille en quinconce, des terres de couleur brune sont visibles. Leur présence n'a pas été reconnue sur les carottages effectués en différents points de l'entrée et du fond de la grotte. Cependant, l'étude couplée des profils d'objets, des coupes stratigraphiques et les observations de terrain permettent de dire que c'est un niveau caractérisé par un apport de matière organique probablement anthropique (présence d'outils). Cette formation aurait ensuite subi une dégradation matérialisée par la présence de nombreuses taches et notamment de phosphates. Certaines poches circulaires font

même penser à des sections d'ossements qui auraient été digérées. Par la suite, l'ensemble aurait été affecté par des déformations avec apparition de fractures. Et enfin, une activité biologique importante se serait développée, les concrétions remplissant les terriers et les galeries des racines n'étant pas affectées par les décrochements.

Enfin, la complémentarité des coupes stratigraphiques et des profils d'objets permet de bien caractériser la poche de décarbonatation. La coupe met en évidence la zone d'action de cette poche. En surface, elle s'étend de la bande C jusqu'à la bande I, puis elle se rétrécit. A l'intérieur de la poche, les phénomènes de décarbonatation et de phosphatogenèse sont nombreux. Il est à noter l'oblitération du litage des sables encore nettement visible dans la partie non affectée par l'altération. Quant aux oxydes métalliques, ils apparaissent sous forme de traînées qui s'enroulent autour des objets, de points, de taches ou de concrétions qui, se déposant principalement au-dessus de l'encroûtement calcaire, soulignent ainsi la base de la poche. Toujours au sein de cette poche, l'absorption des objets est forte. L'étude du profil des objets confirme cette observation. Au niveau de la zone de décarbonatation, les pierres dans F sont quasi absentes, la faune est rare et la présence de galets n'est qu'occasionnelle. Seules les industries lithiques, par leur nature lithologique, ont résisté à l'action géochimique.

Ce travail de description des coupes stratigraphiques, d'étude de leurs recoupements et d'exploitation des profils d'objets est indispensable pour individualiser les niveaux archéologiques et les sols d'occupation qu'ils contiennent, d'en faire une description aussi complète que possible tant sur le plan archéologique que sur le plan sédimentologique, et d'établir la liste du matériel archéologique de chacun des sols. A partir des coordonnées spatiales du matériel archéologique, les topographies actuelles des bases et sommets des différents sols d'habitat peuvent ainsi être représentées sur maillage régulier carré.

Restitution, en plan et en trois dimensions, des divers sols d'habitat préhistorique tels qu'ils devaient se présenter à l'origine, après correction des déformations postdépositionnelles

Depuis sa mise en place, le remplissage quaternaire de la Caune de l'Arago a été affecté par :

- une flexure progressive syngénétique du remplissage alors que les dépôts étaient encore meubles, si bien que les couches les plus basses sont les plus infléchies (en forme de cuvette) et que les couches les plus hautes sont les moins déprimées (en forme d'assiette ou même horizontales pour les plus élevées).

- une évolution géochimique et, en particulier, une paragenèse phosphatée, qui accentue flexure et tassement du remplissage.

- une fracturation et un déplacement de la masse sédimentaire verticalement vers le bas après induration des dépôts due à l'évolution géochimique postdépositionnelle, et suite à la formation d'épais planchers stalagmitiques en surface du remplissage.

• une recarbonatation postérieure et une troncature du remplissage en relation avec l'ouverture de l'aven au plafond de la grotte et l'effondrement du porche originel qui ont favorisé les phénomènes d'infiltration d'eaux de pluie, de ruissellement et de ravinement.

Afin de procéder à la restitution des sols d'habitat tels qu'ils devaient se présenter à l'origine, les corrections des déformations postdépositionnelles doivent être appliquées à la restitution précédente. L'étude des failles et diaclases, la connaissance des caractéristiques sédimentologiques et des qualités plastiques des matériaux du remplissage, et l'analyse de la fracturation du matériel archéologique sont autant d'éléments permettant le calcul de ces déformations. Cette méthode de calcul des déformations postdépositionnelles a été mise en application par A. Fournier[1] pour étudier la flexure et le tassement de l'ensemble stratigraphique III de la Caune de l'Arago.

Les failles et diaclases qui affectent les dépôts quaternaires de la Caune de l'Arago sont nombreuses. L'observation des rejets à leur contact permet de quantifier l'étirement des couches sédimentologiques et donc le taux de déformation qui a affecté le remplissage. Cette déformation est précisément représentée sur les coupes stratigraphiques dont le relevé à l'échelle 1/10ième met en évidence les décalages altitudinaux. Cependant, le développement de la fracturation est plus ou moins net selon la nature du sédiment (Miskovsky, 1988).

Dans le cas des argiles, la rupture ne s'observe que sur un sédiment sec. C'est également le cas des limons. Les sables réagissent de façon différente suivant qu'ils sont fins ou grossiers, lités ou non. Or, sables, limons et argiles sont en quantité variable dans les remplissages de grotte. Il existe tous les intermédiaires entre l'argile pure et le sable pur. La détermination de la proportion relative des différents éléments en présence permet donc d'appréhender le comportement du remplissage sédimentaire de la cavité karstique vis-à-vis de l'eau et de la déformation.

Le matériel archéologique enregistre également très nettement la déformation. Alors que la couche sédimentaire s'étire plus ou moins facilement, l'objet fragile se fracture ou les connexions anatomiques disparaissent. Les deux parties précédemment en connexion s'écartent alors suivant l'orientation de la déformation. C'est ce déplacement qui est étudié.

Par exemple, en juillet 2000, en zone D18, un humérus gauche anténéandertalien (Arago 82) a été mis au jour à la surface du sol G (fig. 5). Brisé en quatre fragments séparés entre eux par quelques millimètres, il témoigne non seulement de l'affaissement des niveaux archéologiques et des couches stratigraphiques mais également de leur fracturation.

La fissuration la plus à l'Ouest ne présente aucune déformation. Les deux parties de la diaphyse sont toujours en contact. La suivante est caractérisée par une déformation Nord-Sud nettement visible sur la vue de dessus de l'humérus. La dernière fracture, la plus à l'Est, sépare de quelques millimètres l'extrémité distale de l'humérus de la diaphyse. La vue de profil montre que les deux parties de l'humérus n'ont pas le même pendage (fig. 6). Cette déformation témoigne de la flexuration de la couche stratigraphique.

APPLICATION

Epais et très dense en matériel archéologique, le sol G est un des niveaux d'habitat les plus caractéristiques du remplissage de la Caune de l'Arago. L'étude de la surface de ce sol est particulièrement intéressante puisqu'une moitié (des bandes 14 (pour un Y compris entre 50 et 100) à 18 (pour un Y compris entre 0 et 50)) a déjà été fouillée dans les années 70 et que l'autre, actuellement en cours de fouilles, est encore visible sur le terrain, dans les bandes 18 (pour un Y compris entre 50 et 100), 19 et 20. Le choix d'étudier le sol G a donc paru évident. L'étude a été menée au niveau des bandes E à I, en 15, 16 et 17, au cœur de la poche décarbonatée (fig. 7).

Figure 5 - Humérus humain gauche en place orientation Ouest / Est, vue supérieure.

Figure 6 - Humérus humain gauche en place orientation Ouest / Est, vue de profil.

(Photothèque du C.E.R.P., Tautavel)

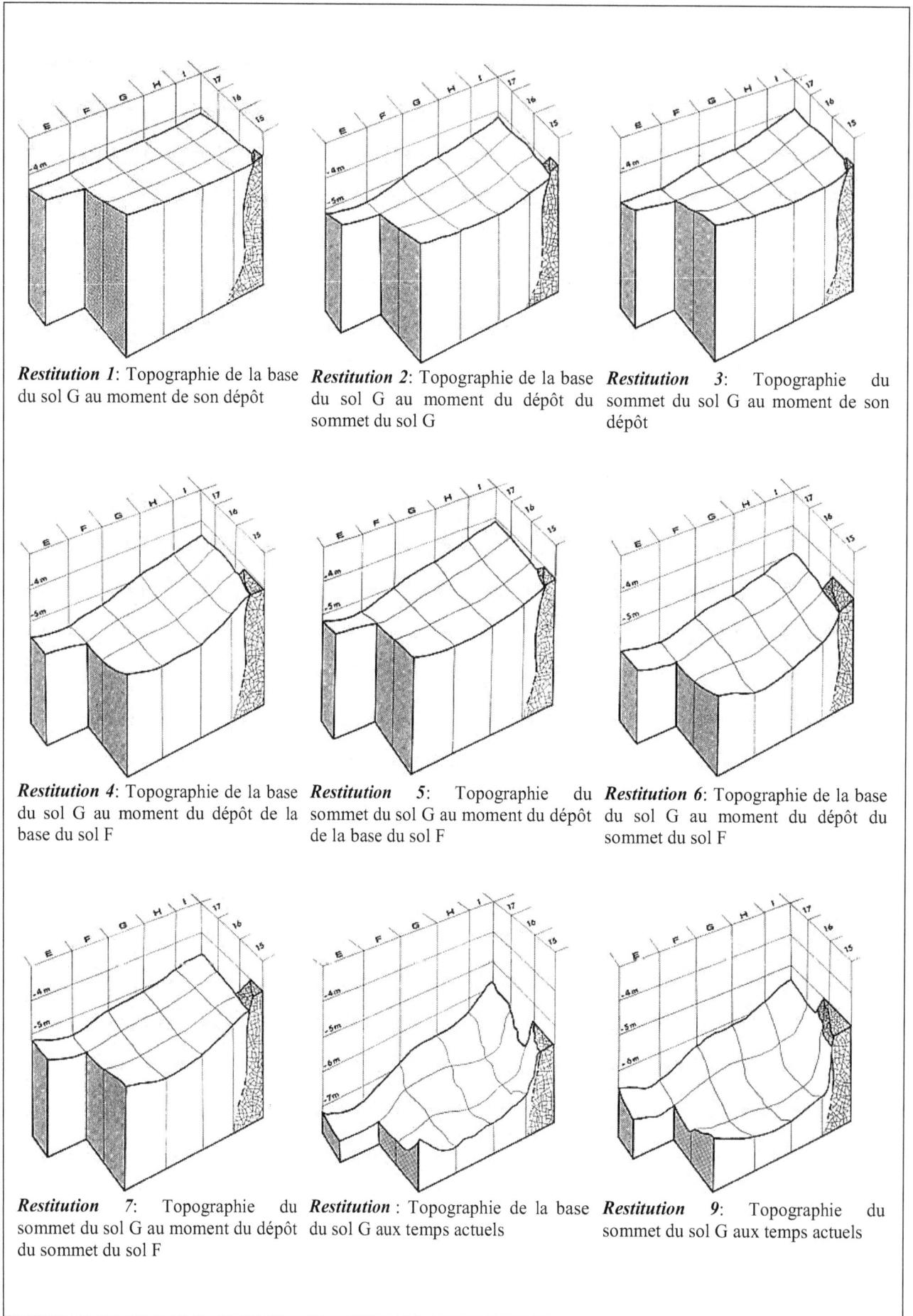

Restitution 1: Topographie de la base du sol G au moment de son dépôt

Restitution 2: Topographie de la base du sol G au moment du dépôt du sommet du sol G

Restitution 3: Topographie du sommet du sol G au moment de son dépôt

Restitution 4: Topographie de la base du sol G au moment du dépôt de la base du sol F

Restitution 5: Topographie du sommet du sol G au moment du dépôt de la base du sol F

Restitution 6: Topographie de la base du sol G au moment du dépôt du sommet du sol F

Restitution 7: Topographie du sommet du sol G au moment du dépôt du sommet du sol F

Restitution : Topographie de la base du sol G aux temps actuels

Restitution 9: Topographie du sommet du sol G aux temps actuels

Figure 7- Reconstitutions des modifications topographiques de la base et du sommet du niveau archéologique G depuis son dépôt jusqu'aux temps actuels, Caune de l'Arago, Tautavel, Pyrénées-Orientales (Butour C., 2001).

Au moment de sa mise en place, le niveau archéologique G ne présente qu'une légère allure en cuvette qui s'accentue alors que se mettent en place les niveaux supérieurs. L'affaissement est plus important dans le centre de la cuvette, en bande E et F, que vers les parois où les couches s'accrochent. Le niveau archéologique G se plaque même verticalement contre la paroi Est de la grotte. Par ailleurs, au cours de la déformation, la dissymétrie de la cuvette se confirme. En 14/15, le point d'enfoncement maximum est situé à la limite des bandes F et G ; en 17/18, ce point est situé en bande E. Enfin, la contemporanéité de l'affaissement et des dépôts de l'ensemble stratigraphique III de la Caune de l'Arago explique que les couches sont d'autant plus flexurées qu'elles sont plus basses, et que les couches situées entre deux niveaux sont plus épaisses dans le centre que vers les parois (Lumley H. de et al., 1981).

CONCLUSION

La fouille d'un site préhistorique met au jour un important matériel archéologique. Chaque campagne de terrain est unique et ne peut être renouvelée puisque le travail de fouille entraîne la destruction du gisement. Les données n'étant donc lues qu'une fois, il est primordial de relever un maximum de renseignements sur le terrain.

De plus, au fur et à mesure des progrès de la discipline, toutes les données archéologiques et les notes d'observation restant valables, elles doivent être prises en compte lors des nouvelles études et interprétations. Employer des procédés manuels pour traiter cette abondante information devient alors très difficile, voire même inefficace. L'informatisation des données archéologiques permet désormais aux préhistoriens de profiter des applications nouvelles que sont la gestion et le traitement des informations la représentation spatiale et la simulation.

Cependant, en raison de la spécificité de chaque gisement, la Préhistoire est un domaine qui n'autorise pas l'utilisation directe de programmes informatiques classiques mais, au contraire, l'élaboration de techniques propres à cette discipline. En fait, le problème n'est pas de savoir quel micro-ordinateur utiliser ou quel logiciel est le meilleur. Il faut savoir quelles données informatiser, comment et pourquoi, et il s'agit donc de définir les éléments à traiter.

La Caune de l'Arago a été choisie pour sa richesse en matériel archéologique et son enregistrement de l'altération et des déformations synsédimentaires et postdépositionnelles qui ont permis de mettre en application cette méthode de travail. Dans le cadre de ce travail, la faisabilité de la méthode a uniquement été testée sur le niveau archéologique G. Cette étude doit donc être poursuivie afin d'appliquer la méthode à l'ensemble des sols d'habitat préhistorique. A partir de ces restitutions en trois dimensions des sols d'habitat préhistoriques, l'évolution géodynamique des dépôts quaternaires pourra être simulée. A travers cette visualisation en 3D, chercheurs et étudiants auront une nouvelle approche de la répartition spatiale de leur matériel. Appliquée à de nombreux autres gisements, la méthode développée rendra possible la reconstitution des paléoenvironnements du Bassin méditerranéen par confrontation des interprétations du remplissages des divers sites.

Adresses des auteurs :

Henry de LUMLEY, Institut de Paléontologie Humaine, 1, rue René Panhard, 75013 Paris, France.

Caroline BUTOUR et Véronique POIS, Centre Européen de Recherches Préhistoriques, UMR 5590 du CNRS ,Avenue Léon Jean Grégory ,66720 Tautavel, France.

Anne-Marie MOIGNE, Laboratoire de Préhistoire du Muséum National d'Histoire Naturelle, UMR 6569 du CNRS, & Centre Européen de Recherches Préhistoriques ,

Rachel VAUDRON, Laboratoire départemental de Préhistoire du Lazaret, 06000 Nice, France.

Note: Alain Fournier, Ingenieur d'Étude au C.N.R.S., attaché a l'Unité Mixte de Rcherche 6569 « L'Homme Préhistorique : son évolution, son milieu, ses activités », à la Faculté des Sciences Saint Charles, à Marseille.

BIBLIOGRAPHIE

BUTOUR, C., 2001, Evolution géodynamique des dépôts quaternaires de la Caune de l'Arago (Tautavel, Pyrénées-Orientales). *Thèse de Doctorat « Quaternaire : Géologie, Paléontologie Humaine, Préhistoire », option Géologie.* Université de Perpignan, 388 p., 49 fig., 4 annexes, 97 fig. en annexe I, 58 réf. bibl.

CANALS i SALOMO, A., 1993, Méthode et technique archéo-stratigraphique pour l'étude des gisements en sédiment homogène : application au complexe CIII de la grotte du Lazaret, Nice (Alpes-Maritimes). Informatique appliquée : base de données et visualisation tridimensionnelle d'ensembles archéologiques. *Thèse du Muséum d'Histoire Naturelle,* Paris, 124 p., 72 fig., 5 tabl., 42 réf. bibl.

COLLECTIF, 1992, Panneaux explicatifs, Musée de la Préhistoire de Tautavel.

COLLECTIF, 1997, Dictionnaire encyclopédique de l'Information et de la documentation, Collection « réf. » aux Editions Nathan 1997.

FRUITET, J., 1991, De l'archéologie préhistorique à l'archéologie numérique : outils informatiques et méthodologie pour une base de données relationnelle de matériel préhistorique et paléontologique. Aide à la détermination de niveaux archéologiques. *Thèse de l'Université de Paris VII*, Paris, 211 p.

LUMLEY, H. de, FOURNIER, A., MISKOVSKY, J.-C., BOUDIN, R -C., PENEAUD, P., BEINER, M., PARK, Y.-C., CAMARA, A., GELEIJNSE, V., SAAS, A., HOFFERT, M., SCHAAF, O. , 1981, Evolution géochimique du remplissage de la Caune de l'Arago à Tautavel, postérieure à la mise en place des sédiments. In *Colloque international du C.N.R.S, Datations absolues et analyses isotopiques en Préhistoire. Méthodes et limites. Datation du remplissage de la Caune de l'Arago, Tautavel, 22-28 juin 1981,* Edited by laboratoire de Préhistoire du Muséum de l'Homme, dir H. de Lumley et J. Labeyrie, 15 fig., p. 79-93.

MISKOVSKY, J.-C. ,1988, Les sédiments, témoins du passé. Editions Science et Découvertes « Le Rocher », 117 p.

POIS, V., 1998, La Caune de l'Arago (Pyrénées-Orientales) : visualisation spatiale, en coupe et en plan, du matériel archéologique par interrogation de la base de données « Matériel Paléontologique et Préhistorique ». Conséquences sur l'interprétation du mode de vie de l'Homme de Tautavel. *Thèse de Doctorat de 3° cycle, Géologie du Quaternaire, Paléontologie Humaine, Préhistoire.* Muséum d'National d'Histoire Naturelle de Paris, 2 tomes, 425 p., 100 fig., annexes I, II et III.

www.ingramcontent.com/pod-product-compliance
Lightning Source LLC
Chambersburg PA
CBHW061008030426
42334CB00033B/3406